Longman Exam Guides
Quantitative Methods

Longman Exam Guides

Series Editors: **Stuart Wall and David Weigall**

Titles available:

Accounting Standards
Bookkeeping and Accounting
British Government and Politics
Business Communication
Business Law
Economics
English as a Foreign Language: Preliminary
English as a Foreign Language: Intermediate
English Literature
French
Monetary Economics
Office Practice and Secretarial Administration
Pure Mathematics
Quantitative Methods
Secretarial Skills

Forthcoming:

Accounting: Cost and Management
 Financial
Biology
Business Studies
Chemistry
Commerce
Computer Science
Electronics
Elements of Banking
English as a Foreign Language: Advanced
General Principles of Law
General Studies
Geography
Mechanics
Modern British History
Physics
Sociology
Taxation

Longman Exam Guides

QUANTITATIVE METHODS

D. V. Friend, M.Sc., M.Sc. (Econ.)

LONGMAN
London and New York

Longman Group UK Limited
Longman House, Burnt Mill, Harlow,
Essex CM20 2JE, England
and Associated companies throughout the world

*Published in the United States of America
by Longman Inc., New York*

© Longman Group UK Limited 1987

First published 1987

British Library Cataloguing in Publication Data

Friend, D.V.
  Quantitative methods.—(Longman exam guides)
  1. Business mathematics
  I. Title   II. Series
  510'.24658   HF5691

ISBN 0-582-29696-X

Library of Congress Cataloging in Publication Data

Friend, D.V. (Derek V.), 1927–
  Quantitative methods.
  (Longman exam guides)
  Includes index.
  1. Mathematics—Problems, exercises, etc.
I. Title.   II. Series.
QA43.F69   1986   510'.76   86–7166
ISBN 0–582–29696–X

Set in 9½ on 11 pt Lasercomp Times Roman by August Filmsetting, Haydock

Produced by Longman Singapore Publishers (Pte) Ltd.
Printed in Singapore.

Contents

Editors' Preface

Much has been said in recent years about declining standards and disappointing examination results. Whilst this may be somewhat exaggerated, examiners are well aware that the performance of many candidates falls well short of their potential. Longman Exam Guides are written by experienced examiners and teachers, and aim to give you the best possible foundation for examination success. There is no attempt to cut corners. The books encourage thorough study and a full understanding of the concepts involved and should be seen as course companions and study guides to be used throughout the year. Examiners are in no doubt that structured approach in preparing for and taking examinations can, together with hard work and diligent application, substantially improve performance.

The largely self-contained nature of each chapter gives the book a useful degree of flexibility. After starting with Chapters 1 and 2, all other chapters can be read selectively, in any order appropriate to the stage you have reached in your course. We believe that this book, and the series as a whole, will help you establish a solid platform of basic knowledge and examination technique on which to build.

Stuart Wall and David Weigall

Author's Preface

At the Polytechnic of Central London we prepare students for a variety of examinations in Business Studies and the Social Sciences: BA Business Studies and BA Social Sciences, both full-time and part-time, HND in Business Studies, Accounting Foundation Course, Institute of Cost and Management Accountants, Institute of Chartered Secretaries and Administrators and Chartered Association of Certified Accountants. For many of these courses, particularly in the evening school, we have to employ part-time lecturers. To help both evening and full-time staff I developed a series of handouts and exercise sheets. Over the years these handouts have been amended and improved as a result of comments by students and colleagues. I should like to express my thanks for this help.

The text and many of the worked examples and exercises were based on these handouts and exercise sheets, the remainder of the worked examples and exercises were taken from recent examination questions set by professional bodies. I should like to thank my son, Graham, for checking the answers to the worked examples and the exercises on statistical topics. I should also like to thank my wife, Jennifer, and my daughter, Stephanie, for checking the manuscript and ensuring that the spelling and grammar were correct.

I should also like to express my thanks to Stuart Wall for the patience and help in clarifying and improving the text.

Acknowledgements

We are indebted to Marks and Spencer plc for permission to reproduce extracts from the Company's 1985 *Report and Financial Statements Booklet*, and the following for questions from past examination papers:
The Association of Business Executives; The Chartered Association of Certified Accountants; The Institute of Chartered Secretaries and Administrators; The Institute of Cost and Management Accountants; The Royal Society of Arts Examination Board.

The solutions to examination questions are entirely my responsibility and have neither been provided nor approved by the examination bodies.

Greek letters

In this book some Greek letters have been used. They are as follows:

μ = Greek letter mu
Σ = Greek letter sigma upper case
σ = Greek letter sigma lower case
π = Greek letter pi

Chapter 1

Coverage of topics

Most professional bodies include a compulsory quantitative element at the introductory level. Similarly most first-year courses for degrees and diplomas in Accounting, Business Studies, Management Studies and Social Sciences have a compulsory module in the quantitative area. Until fairly recently the quantitative element was mostly statistics, but now most courses include some mathematics. The mathematical content varies but usually includes algebra, differential calculus and financial mathematics. Some courses include an introduction to certain operational research methods, the most common being linear programming.

This book aims to cover most of the introductory syllabuses in the quantitative area, and most of the standard questions set. It does not attempt to cover topics in the *later* stages of professional, B/TEC or undergraduate courses in any depth. However, if you do use this book in the early part of such courses, you will find that it will be of some use in higher-level units later on.

Table 1.1 on p. 2 shows the relevance of the topic areas for the various examinations. The columns refer to:

ICMA	Institute of Cost and Management Accountants
CACA	Chartered Association of Certified Accountants
ICSA	Institute of Chartered Secretaries and Administrators
ABE	Association of Business Executives
LCCI	London Chamber of Commerce and Industry
RSA	Royal Society of Arts
OND/C	Ordinary National Diploma/Certificate in Business Studies
HND/C	Higher National Diploma/Certificate in Business Studies
DMS	Diploma in Management Studies
IM	Institute of Marketing

Table 1.1

Chapter and topic	AAT — Numeracy and Statistics (Prelim.)	AAT — Economics and Statistics (Interm.)	ABE — Bus. Stats.	BTEC — OND/C App. Stats.	BTEC — HND/C Quant. Mthds.	CACA (ACCA) — Num. Anal. &DP	CII — QM (Qual.)	DMS — QM	ICA (E8W) — Quant. Techn. (Foundation)	ICA (Ireland) — Statistics (Prof. 1)	ICA (Scotland) — Math. Techn. (Prel.)	ICM — Bus. Stats.
3. Revision of arithmetic and algebra	✓	✓	✓	✓	✓	✓	✓	✓	✓	✓	✓	✓
4. Presentation of data	✓	✓	✓	✓	✓	✓	✓	✓	✓	✓	✓	✓
5. Measures of location and dispersion	✓	✓	✓	✓	✓	✓	✓	✓	✓	✓	✓	✓
6. Regression and correlation		✓	✓	✓	✓		✓	✓	✓	✓	✓	✓
7. Index numbers	✓	✓	✓	✓	✓	✓	✓	✓	✓	✓	✓	✓
8. Time series		✓	✓	✓	✓	✓	✓	✓		✓	✓	✓
9. Probability		✓	✓	✓	✓			✓	✓	✓	✓	✓
10. Normal, binomial and Poisson distributions		✓	✓		✓		✓	✓	✓	✓	✓	✓
11. Confidence intervals and significance testing		✓	✓	✓	✓		✓	✓	✓	✓	✓	
12. Algebra applied to business and economics	✓				✓	✓	✓	✓	✓	✓	✓	
13. Calculus					✓	✓		✓	✓	✓		
14. Financial mathematics	✓			✓	✓	✓		✓	✓		✓	
15. Linear programming					✓			✓	✓		✓	

Course	ICMA	ICSA	IM	IPS	LCCI		RSA		SCOTVEC	SCCA
Chapter and topic	Quant. Methods (Stage 1)	Quant. Stud. (Part 1)	Quant. Stud. (Part 1)	Quant. Stud. (Foundation)	Bus. Stats. (Inter.)	Bus. Stats. (Higher)	Statistics (Stage I)	Statistics (Stage II)	HND/C	Stat. Method (Part 2)
3. Revision of arithmetic and algebra	✓	✓	✓	✓	✓	✓	✓	✓	✓	✓
4. Presentation of data	✓	✓	✓	✓	✓	✓	✓	✓	✓	✓
5. Measures of location and dispersion	✓	✓	✓	✓	✓	✓	✓	✓	✓	✓
6. Regression and correlation	✓	✓	✓	✓	✓	✓	✓	✓	✓	✓
7. Index numbers	✓	✓	✓	✓	✓	✓	✓	✓	✓	✓
8. Time series	✓	✓	✓	✓	✓	✓		✓	✓	✓
9. Probability	✓	✓	✓	✓	✓	✓			✓	✓
10. Normal, binomial and Poisson distributions	✓	✓	✓	✓	✓	✓			✓	✓
11. Confidence intervals and significance testing	✓	✓	✓	✓		✓			✓	✓
12. Albegra applied to business and economics	✓	✓	✓	✓					✓	
13. Calculus	✓									
14. Financial mathematics	✓	✓	✓	✓					✓	
15. Linear programming	✓	✓	✓	✓						

To use this book you should obtain the syllabus of the examination for which you are studying. In the case of professional bodies this is issued to students. In the case of college courses which are internally examined, the syllabus is not necessarily a good guide to the topics set in examination questions. As an external examiner and as an assessor I have noticed that, at some colleges, some parts of the syllabus are obviously never taught, and some topics are included which are not in the syllabus. On this type of college course you will normally be given a scheme of work for the academic year: this will be your best guide to the syllabus. In practice most college lecturers set questions only on the topics actually taught, thus your lecture notes are a very good indicator of likely examination questions.

On p. 6 it is suggested that you obtain past examination papers and make an analysis of the questions set. In this way you can select the chapters of this book which need to be studied in most depth. Some guidance is given on how best to use this book on p. 8 under the heading 'passing the examination.'

FURTHER READING

This book is written assuming that the reader has, at best, 'O'-level mathematics and that most of the mathematics once learned has been forgotten. There will however be some readers who are numerate, having followed a course in 'A'-level mathematics or obtained an equivalent qualification. Such readers may prefer, in addition, a more mathematical approach. The following books are suggested as appropriate further reading:

Fundamentals of Statistics Mulholland and Jones. Heinemann, 1968.
Mathematics for the Manager Tennant-Smith. Van Nostrand Reinhold, 1973.

At the end of most chapters there is a section 'A step further', in which appropriate reading from these books is suggested.

Chapter 2   Examination techniques

Most examination questions in quantitative methods require a candidate to manipulate data, produce results and to comment on the results. Quantitative questions of this type are the major concern of this book. In some examinations, one or two questions are set which require 'essay type' answers – these questions usually cover survey methodology and sources of social, business and economic statistics. On p. 8 appropriate reading is suggested for students wishing to cover survey methodology and sources of statistics.

MATHEMATICAL BACKGROUND

The major accounting bodies require students to hold 'O'-level mathematics or an equivalent qualification, although other bodies, such as the Institute of Chartered Secretaries and Administrators and the Association of Business Executives, do not have such requirements. The problem in quantitative subjects is that usually two or more years have elapsed since mathematics was studied at school, and, in that time, students have often forgotten most of the mathematical topics. It is not necessary to revise *all* the 'O'-level syllabus, but certain topics must be revised. If the examination you are taking contains questions on statistics only, then fewer mathematical topics are needed. Revision of arithmetic and algebra is covered in Chapter 3 and it is strongly recommended that you study Chapter 3 first if your mathematics is rusty.

Many students are very worried about quantitative methods; their memories of school mathematics make them think the subject is going to be difficult. However, most of the topics covered in the quantitative area are clearly relevant to business accounting and the social sciences. If the student can see a use for the subject, then its study becomes more attractive. Some students who start off with a low level of numeracy are very pleasantly surprised how well they cope with quantitative subjects.

CALCULATORS

Almost all examination bodies allow the use of calculators. Examiners often set questions on the assumption that all candidates will have calculators, so that it is essential to have a calculator. If the examination paper covers statistics only, then a calculator with a square-root facility is adequate. If the examination paper involves mathematics, particularly financial mathematics, then a scientific calculator is preferable (for 'O'-level mathematics most candidates buy a scientific calculator).

Most examination bodies will not allow programmable calculators but some calculators, which are allowed, are pre-programmed to work out certain results. The usual requirement expressed in the rubric of the examination paper is that appropriate intermediate steps *must be shown*. This means that if your calculator will (say) find the correlation coefficient automatically, then, if you use this facility and do not show intermediate steps, you could be very heavily penalized. If, however, you work out the correlation coefficient showing intermediate steps and then *check* the accuracy by using the automatic facility, the examiner will never know! You should find out the rules on calculators set out in the examination regulations and read the rubric of past examination papers.

PAST EXAMINATION PAPERS

The syllabus sets out the topics to be covered: the examination papers show how the syllabus is interpreted. It is essential to obtain recent, past examination papers. Some examination bodies publish full answers to examination papers. These are very helpful, particularly for final revision. You can attempt a past paper under examination conditions and then check your answers.

Past examination papers show how many questions are to be attempted and the choice of questions. The number of questions to be attempted is usually between four and six. The number of questions set varies. For the Chartered Association of Certified Accountants candidates are required to attempt five questions out of six, so the choice is very limited. Five questions out of eight is more common.

It is worth making a detailed analysis of the last five or six past examination papers and then setting out in tabular form the topics covered. Construct a grid similar to that shown in Chapter 1 (Table 1.1) but put *years* along the top of the page and put a tick against each topic. You may be able to predict possible questions! You will certainly be able to identify the topics which are set most frequently, and therefore the topics to be studied most intensively.

FORMULAE LIST

Some examination bodies supply a formulae list in the examination room (often it is attached to the actual paper). If a formulae list is supplied you need to examine the list carefully. There are different formulae for a particular measure in the various textbooks, so you must make certain that you know how to use the *formulae given in the formulae list*. Any formulae not given in the list have to be learned. If a formulae list is not supplied, you have to learn all the relevant formulae.

TABLES

Check to see if any tables are supplied (they sometimes form part of the formulae list). If tables are supplied, check to see if you know how to use them. This book uses the normal distribution tables showing the area to the left of the ordinate. Some examination bodies use tables showing the area to the right of the ordinate.

EXAMINATION ROOM STRATEGY

In college-set examinations the pass mark is usually 40% and as the college lecturers set the examinations you will normally have been taught all the topics examined. For most professional examinations the pass mark is usually 50%, and the college lecturers have no knowledge of the content of the paper. This makes passing the examinations of the professional bodies more difficult.

If you are taking any examination you must use your time in the examination room effectively. In an examination which requires essays, you have to complete the essay because the examiner expects a balanced answer and a conclusion. In quantitative subjects you score marks *for each appropriate step*. The marks are divided into 'method' marks and 'accuracy' marks. Method marks are awarded for following the correct method, and accuracy marks are awarded for correct arithmetic or algebra. To obtain accuracy marks the method has to be correct. If you make an early arithmetic error in a question, you are not usually penalized for consequent arithmetic errors provided the method is correct.

When attempting a paper you must start by doing the questions where you are certain of the method. If you have five questions to do in three hours, you have 35 minutes per question. Never spend more than 35 minutes on a question. Usually in the earlier parts of questions the 'method' and 'accuracy' marks are easier to score. After 35 minutes on a question you will probably find marks easier to score on *the next question*. If you have time at the end of the examination, you can go back and try to finish incomplete questions.

Finally, when you have obtained a numerical answer to a question, ask yourself 'Is the answer realistic?'. As an examiner I have seen the mean age of a mother given as 248 years! This is obviously wrong and a quick check may reveal the source of the error. A correlation coefficient must lie between -1 and $+1$, and any answer numerically greater than 1 is again obviously wrong.

EXAMINATION TOOLS

An examination in quantitative methods usually requires some graphical work. For instance, you could be expected to draw a pie chart. Far too many candidates go into the examination room without the tools for the job. Make certain you have the following:

calculator (and if possible a spare calculator)	pencil sharpener
batteries for calculator	rubber
ruler	protractor
pencils (perhaps coloured pencils as well)	pair of compasses

PASSING THE EXAMINATION

The only way to pass an examination in quantitative subjects is to attempt successfully a large number of exercises *before* you take the examination. This book has many worked examples, and after each worked example there is an exercise. Try the exercise in order to make certain that you understand the worked example. The answer is given at the end of the question. If you do not obtain the correct answer, check the method and the arithmetical accuracy of your working. If you *still* cannot obtain the correct answer, check the solution presented under the section heading C 'Solutions to Exercises'. When you have finished the chapter, attempt the questions taken from recent examination papers in section D *before* checking your answers and methods against the Outline Answers presented in section E.

SURVEY METHODOLOGY

The standard work on survey methodology in the United Kingdom is *Survey Methods in Social Investigation* by C. A. Moser and G. Kalton published by Heinemann. This book is very comprehensive, and the mathematical parts can be omitted. This book is stocked by most college and public libraries.

SOURCES OF SOCIAL, BUSINESS AND ECONOMIC STATISTICS

This is not an easy subject to study, since the way the statistics are compiled is constantly changing. Most books on the subject are out of date when they are first published. The most recent books in this area are published by the Open University.

The Central Statistical Office has written *A Guide to Official Sources* published by HM Stationery Office. This book lists most of the published sources of statistics and is stocked by many libraries. *UK Statistical Sources*, edited by Professor P. Maunder and published by Pergamon, is also stocked in many libraries. This is a comprehensive series of individual books, each one giving detailed information on the statistical sources of a *particular industrial sector*. Most candidates in examinations in the quantitative area (perhaps wisely!) avoid questions on sources of statistics.

Chapter 3

Revision of arithmetic and algebra

A. GETTING STARTED

Most students reading this book will have forgotten a considerable amount of the mathematics they once knew. It is essential to make certain that you become proficient again at the elementary arithmetic and algebraic processes.

Examination boards usually allow candidates to have calculators. There will therefore be no revision of basic arithmetic – addition, subtraction, multiplication and division. If you do not own a calculator, buy a *scientific* calculator. If you have a calculator which does not have many functions, then still buy a scientific calculator. It is strongly recommended that a calculator is used for all arithmetic calculations – even multiplying by 2!

B. ESSENTIAL PRINCIPLES

RULES FOR MULTIPLICATION

(a) Like signs

(i) $+ \times + = +$ e.g. $+4 \times +5 = +20$
(ii) $- \times - = +$ e.g. $-3 \times -7 = +21$

(b) Unlike signs

(i) $- \times + = -$ e.g. $-6 \times +7 = -42$
(ii) $+ \times - = -$ e.g. $+3 \times -8 = -24$

RULES FOR INDICES

In the expressions a^5, a^2; a is called the *base*, 5 and 2 are called the *indices*.
$$a^5 = a \times a \times a \times a \times a \qquad a^2 = a \times a$$

(a) Multiplication

$a^5 \times a^2 = a \times a \times a \times a \times a \times a \times a = a^7$
Rule – When we *multiply* we *add* the indices, i.e. $a^5 \times a^2 = a^{(5+2)} = a^7$

(b) Division

$$\frac{a^5}{a^2} = \frac{a \times a \times a \times a \times a}{a \times a} = a \times a \times a = a^3$$

Rule – When we *divide* we *subtract* the indices, i.e.

$$\frac{a^5}{a^2} = a^{(5-2)} = a^3$$

(c) $a^0 = 1$

This may be demonstrated by applying the division rule

$$\frac{a^5}{a^5} = a^{(5-5)} = a^0$$

Now

$$\frac{a^5}{a^5} = \frac{a \times a \times a \times a \times a}{a \times a \times a \times a \times a} = 1 \qquad \text{Thus } a^0 = 1$$

(d) Negative indices

If we divide a^2 by a^5 then by the rule for division

$$\frac{a^2}{a^5} = a^{(2-5)} = a^{-3}.$$

Now

$$\frac{a^2}{a^5} = \frac{a \times a}{a \times a \times a \times a \times a} = \frac{1}{a \times a \times a} = \frac{1}{a^3}$$

Thus

$$a^{-3} = \frac{1}{a^3} \text{ similarly } a^{-4} = \frac{1}{a^4}.$$

In *general*

$$a^{-n} = \frac{1}{a^n}.$$

(e) Indices raised to a power

Suppose we wish to find the value of a^2 cubed, i.e.

$$(a^2)^3$$

Now

$$(a^2)^3 = a^2 \times a^2 \times a^2 = a^{(2+2+2)} = a^6$$

applying the multiplication rule.
Thus

$$(a^2)^3 = a^{(2 \times 3)} = a^6$$

Rule – When indices are raised to a *power* we *multiply* the indices.

(f) Fractional indices

If we square $a^{\frac{1}{2}}$, i.e. $(a^{\frac{1}{2}})^2$ we obtain, by the rule above, $a^{(\frac{1}{2} \times 2)} = a^1$
Now $a^1 = a$, thus $a^{\frac{1}{2}}$ squared $= a$
Therefore $a^{\frac{1}{2}}$ is the square root of a, i.e. $a^{\frac{1}{2}} = \sqrt{a}$

Similarly $a^{\frac{1}{3}}$ is the cube root of a, i.e. $a^{\frac{1}{3}} = \sqrt[3]{a}$

In general $a^{1/n}$ is the nth root of a, i.e. $a^{1/n} = \sqrt[n]{a}$

Worked Example 1

Simplify the following expressions:

(i) $3X^2Y^3 \times 4X^4Y$;   (ii) $\dfrac{2X^2Z^3}{5} \div \dfrac{4X}{15Z^2}$

(i) Collect terms:

$$3 \times 4 \times X^2 \times X^4 \times Y^3 \times Y = 12 \times X^{(2+4)} \times Y^{(3+1)} = 12X^6Y^4$$

(ii) Start by *inverting* the second fraction (i.e. turning it on its 'head') and then multiply. Next, remember the rule for division of fractions

$$\dfrac{2X^2Z^3}{5} \times \dfrac{15Z^2}{4X} = \dfrac{3}{2} \times X^2 \times X^{-1} \times Z^3 \times Z^2 = \dfrac{3}{2}X^{(2-1)}Z^{(3+2)} = \dfrac{3}{2}XZ^5$$

Exercise 1

Simplify:

(i) $2a^3b^4 \times 7ab^3$;   (ii) $\dfrac{9a^2bc^3}{b^2} \div \dfrac{18a^2c^2}{b^3}$

(*Answers:* $14a^4b^7$; $\dfrac{b^2c}{2}$)

RULES OF ARITHMETIC

When expressions contain brackets, indices, multiplication, etc., the following procedure MUST be followed.

First priority – **Brackets** (when there are brackets *within* brackets deal with the innermost bracket first)

Second priority – **Indices**

Third priority – **Multiplication** and/or **division**

Fourth priority – **Addition** and/or **subtraction**

Worked Example 2

If $a = 24$, $b = 13$, $c = 3$, $d = -4$, $n = 10$, $r = 2$, find the value of the following expressions:

(i) $a - r(b - c^2)$; (ii) $\dfrac{n}{2}(2a + (n-1)d)$; (iii) $\dfrac{a(r^n - 1)}{r - 1}$

(i) Substituting, $24 - 2(13 - 3^2) = 24 - 2(13 - 9) = 24 - 2 \times 4 = 24 - 8$
$= 16$

(ii) Substituting, $\dfrac{10}{2}(2 \times 24 + (10 - 1) \times -4) = \dfrac{10}{2}(48 + 9 \times -4)$

$= 5(48 - 36) = 5 \times 12 = 60$

(iii) Substituting, $24\dfrac{(2^{10} - 1)}{2 - 1} = 24\dfrac{(1{,}024 - 1)}{1} = 24 \times 1{,}023 = 24{,}552$

If $a = 2$, $b = -8$, $c = -24$, $d = 12$, $n = 10$, find the values of:

(i) $ad - an + bc$; (ii) $(a-b)(b^2 - 2cd)$; (iii) $\dfrac{-b + \sqrt{b^2 - 4ac}}{2a}$

(*Answers:* 196; 6,400; 6)

PERCENTAGES

Worked Example 3

(i) An article costs £25 without VAT; VAT at 15% is added; find the VAT and the price with VAT; (ii) An article costs £207 with VAT included, find the price without VAT.

(i) To find the value corresponding to a percentage, multiply the amount by the percentage and divide by 100. Thus

$$VAT = 25 \times \frac{15}{100} = £3.75.$$

Price with VAT is £28.75. **Note:**

$$\frac{15}{100} = 0.15,$$

so we could have found £28.75 by multiplying £25 by 1.15.

(ii) Let A be amount without VAT, then $A \times 1.15 = 207$, thus

$$A = \frac{207}{1.15} = £180.$$

Exercise 3

A company gives its employees a 7.5% rise. One employee earns £150 per week; what will be his earnings after the rise? Another employee earns £172 after the rise; what was his pay before the rise? (*Answers:* £161.25; £160)

RATIOS

Worked Example 4

A company offers a special incentive bonus of £5,000 to its sales representatives. This bonus is divided between the three sales representatives with the highest sales in the ratio of their sales. If their sales were £180,000, £120,000 and £100,000, how much bonus would each receive?

We need to divide 5,000 in the ratio 18:12:10. The first step is to add 18, 12 and 10 = 40; there are 40 parts, one part equals 5,000 divided by 40 = 125. The first representative will receive 18 parts = 125×18 = £2,250; the second will receive $125 \times 12 = £1,500$; the third 125×10 = £1,250.

Exercise 4

Three men A, B and C invest £15,000, £11,000 and £10,000 in a company. In the first year the company makes £9,000 profits. It is decided to put 20% of the profits into a reserve and divide the remainder between A, B and C in the ratio of their investments. Find how much is placed in reserve and how much each man receives. (*Answers:* £1,800; £3,000; £2,200; £2,000)

SIMPLE EQUATIONS

Worked Example 5

Solve the following equations:

(i) $2x+3=9-x$; (ii) $5(x-3)=6-2(2x-3)$; (iii) $\dfrac{2x}{3}=x+\dfrac{x-4}{2}-3$

(i) The method is to get x's on one side of the equation and numbers on the other side. Remember the rule that if a quantity is moved from one side of the equation to the other side the sign changes.

$$2x+3=9-x \ \therefore \ 2x+x=9-3 \ \therefore \ 3x=6 \ \therefore \ x=\frac{6}{3}=2$$

(ii) For this type of problem the first step is to remove brackets. Remember: (a) that the number outside the bracket multiplies ALL the terms inside the bracket; (b) if there is a **minus** sign before the bracket then, when the bracket is removed, the signs of ALL the terms inside the bracket change.

$$5(x-3)=6-2(2x-3) \ \therefore \ 5x-15=6-4x+6$$
$$\therefore \ 5x+4x=15+6+6 \ \therefore \ 9x=27 \ \therefore \ x=3$$

(iii) When there are fractions the first step is to remove fractions – in this example multiplying by $2\times3=6$ achieves this. Remember to multiply ALL the terms by 6.

$$\frac{2x}{3}=x+\frac{x-4}{2}-3 \ \therefore \ 6\times\frac{2x}{3}=6x+\frac{6(x-4)}{2}-18$$
$$\therefore \ 4x=6x+3(x-4)-18 \ \therefore \ 4x=6x+3x-12-18$$
$$\therefore \ 12+18=6x+3x-4x \ \therefore \ 30=5x \ \therefore \ x=6$$

It is usual to check whether your solution is correct. In example (ii) above put $x=3$ and see if the right-hand side of the equation is equal to the left-hand side:

$$5(x-3)=5(3-3)=5\times0=0$$
$$6-2(2x-3)=6-2(2\times3-3)=6-2(6-3)=6-2\times3=6-6=0$$

As the left-hand side and the right-hand side both equal 0, this checks that $x=3$ is the solution to the equation.

Exercise 5

Solve the following equations

(i) $3x+5=14$; (ii) $3-4x=6x-7$;

(iii) $5 - 3(x - 4) = 2(3 - 2x)$; (iv) $\dfrac{2x + 3}{5} - \dfrac{x - 2}{2} - \dfrac{2}{3} = \dfrac{2x}{15}$

(*Answers:* 3; 1; -11; 4)

SIMULTANEOUS EQUATIONS

Worked Example 6

Solve the following equations:

(i) $\begin{aligned} x + 2y &= 10 \\ x + \ y &= \ 7 \end{aligned}$ (ii) $\begin{aligned} 2x + 3y &= 16 \\ 3x - \ y &= 13 \end{aligned}$ (iii) $\begin{aligned} 5x + 3y &= 16 \\ 2x + 7y &= 31 \end{aligned}$

There are various methods available to solve this type of problem. Some students are taught at school to use matrix methods. My experience as a teacher and as an examiner is that students are not very successful at solving simultaneous equations using matrices. However, if you are directed in an examination to use matrices to solve simultaneous equations you have no alternative – you will probably get no marks if you use other methods. (Matrix algebra is covered later in this chapter.) Students are more successful with the elimination method which is used in the following solutions.

(i) $x + 2y = 10$ ... (1)
    $x + \ y = \ 7$ ... (2)

In order to eliminate x subtract the second equation from the first equation.

$$x - x + 2y - y = 10 - 7 \ \therefore \ y = 3$$

To find x substitute $y = 3$ in one of the equations.

$$x + y = 7 \ \therefore \ x + 3 = 7 \ \therefore \ x = 7 - 3 \ \therefore \ x = 4$$

Check the result by substituting $y = 3$, $x = 4$ in the other equation, i.e. $x + 2y = 10$, $4 + 2 \times 3 = 10$.

(ii) $2x + 3y = 16$ ... (1)
    $3x - \ y = 13$ ... (2)

To effect the elimination process it is necessary for the coefficients of x (or y) to have the same numerical value. To achieve this multiply all the terms of the second equation by 3. The equations become

$$\begin{aligned} 2x + 3y &= 16 \\ 9x - 3y &= 39 \end{aligned}$$

As the signs of the coefficients of y differ, the method of eliminating y is to add the two equations.

$$2x + 9x + 3y - 3y = 16 + 39 \ \therefore \ 11x = 55 \ \therefore \ x = 5$$

To find y put $x = 5$ in one of the equations.

$$\begin{aligned} 2x + 3y = 16 \quad &\therefore \ 2 \times 5 + 3y = 16 \quad \therefore \ 10 + 3y = 16 \\ \therefore 3y = 16 - 10 \quad &\therefore \ 3y = 6 \quad \therefore \ y = 2 \end{aligned}$$

Again, check the result in the other equation; $3 \times 5 - 2 = 13$.

(iii) $5x + 3y = 16$ . . . (1)

     $2x - 7y = 31$ . . . (2)

In this example it is necessary to multiply the first equation by 2 and the second equation by 5

$$10x + 6y = 32$$
$$10x - 35y = 155$$

We now subtract the second equation from the first equation.

$$10x - 10x + 6y - (-35y) = 32 - 155 \qquad \therefore \ 6y + 35y = -123$$
$$\therefore \ 41y = -123 \qquad \therefore \ y = -3$$

To find x put $y = -3$ in one of the equations.

$$5x + 3 \times (-3) = 16 \qquad \therefore \ 5x - 9 = 16$$
$$\therefore \ 5x = 16 + 9 \qquad \therefore \ 5x = 25 \qquad \therefore \ x = 5$$

Again, check the result in the other equation; $2 \times 5 - 7 \times (-3) = 31$.

Exercise 6

Solve the following equations:

(i) $\begin{aligned} 2x + 3y &= 22 \\ 5x - 3y &= 13 \end{aligned}$ (ii) $\begin{aligned} 3x - 4y &= 4 \\ x + y &= 6 \end{aligned}$ (iii) $\begin{aligned} 6x - 5y &= 45 \\ 4x - 7y &= 41 \end{aligned}$

(*Answers:* $x = 5 \ y = 4$; $x = 4 \ y = 2$; $x = 5 \ y = -3$)

QUADRATIC EQUATIONS

Worked Example 7

Solve the following equations:

(i) $x^2 + 7x + 12 = 0$; (ii) $x^2 - 3x - 10 = 0$; (iii) $2x^2 - 3x - 6 = 0$

Equations (i) and (ii) can be solved by factorizing, equation (iii) can be solved either by the 'completing the square' method or by using a formula. Most students are not very successful with the 'completing the square' method, and this method is therefore not recommended. When solving quadratic equations use the formula unless the factors can be found easily.

(i) $x^2 + 7x + 12 = 0$

The first step is to find two numbers such that their product is equal to the constant term, in this case $+12$, and their sum is equal to the coefficient of x, in this case $+7$.

Product of two numbers $= +12$, sum of two numbers $= +7$

By trying the factors of $+12$ we find $+3$ and $+4$ are the numbers required.

$$x^2 + 7x + 12 = (x + 3)(x + 4)$$

(as a revision of multiplying brackets check that

$$(x + 3)(x + 4) = x^2 + 7x + 12)$$

Thus

$$(x+3)(x+4)=0$$

For a product to equal zero, one of the terms of the product must be zero.

$$x+3=0 \quad \text{or} \quad x+4=0$$
$$\therefore x=-3 \qquad \qquad x=-4$$

are the solutions to the equation.

(ii) $x^2-3x-10=0$

Product of two numbers $=-10$. Sum of two numbers $=-3$.
We find that -5, $+2$ satisfy these conditions.

$$(x-5)(x+2)=0$$
$$x-5=0 \quad \text{or} \quad x+2=0$$
$$\therefore x=5 \quad \text{or} \qquad x=-2$$

(iii) $2x^2-3x-6=0$

This does not factorize, so we need to use the formula.
If $ax^2+bx+c=0$ is the general quadratic equation we may show that

$$x=\frac{-b\pm\sqrt{b^2-4ac}}{2a}$$

provides the two solutions ($\pm$ is the sign which means $+$ or $-$).
Comparing $2x^2-3x-6=0$ with the general equation $ax^2+bx+c=0$,
$a=2$, $b=-3$, $c=-6$.
Substituting in the formula,

$$x=\frac{-(-3)\pm\sqrt{(-3)^2-4\times2\times(-6)}}{2\times2}$$

$$x=\frac{3\pm\sqrt{9+48}}{4} \quad \therefore \quad x=\frac{3\pm\sqrt{57}}{4} \quad \therefore \quad x=\frac{3\pm7.5498}{4}$$

$$x=\frac{3+7.54984}{4}=\frac{10.5498}{4}=2.64$$

or

$$x=\frac{3-7.54984}{4}=\frac{-4.5498}{4}=-1.14$$

Exercise 7

Solve the following equations:

(i) $x^2-11x+24=0$; (ii) $x^2-4x-45=0$; (iii) $3x^2+5x-4=0$

(*Answers:* 3, 8; 9, -5; 0.591, -2.257)

GRAPHS

If we are asked to plot the point $x = +2$, $y = +3$ on graph paper the first step is to draw the axes. It is usual to have the x axis horizontal and the y axis vertical. We have to choose appropriate scales. If the values of x and y to be plotted are small you can use the scale in Fig. 3.1, but if the values of x and y are large the scale may have to be 10 20 30 40 or even 100 200 300.

The point $x = +2$, $y = +3$ is shown as point (a) on Fig. 3.1. Point (a) is at the *intersection* of the vertical line from the point $x = +2$ on the x axis and the horizontal line from the point $y = +3$ on the y axis. The following points are also plotted on Fig. 3.1: (b) $x = +3.5$, $y = -2$; (c) $x = -4$, $y = +2$; (d) $x = -2.5$, $y = -1.5$; (e) $x = +4$, $y = 0$.

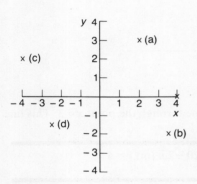

Fig. 3.1

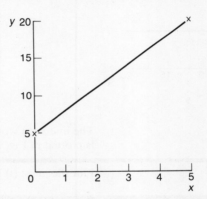

Fig. 3.2

PLOTTING A STRAIGHT LINE

The equation of a straight line is of the form $y = mx + c$, for example $y = 3x + 5$. If we wish to plot a straight line on graph paper, we need to find two points on the line and plot these two points and then join the two points using a ruler.

Worked Example 8

Plot the line $y = 3x + 5$.

To plot the line $y = 3x + 5$ take two x values – you can choose any values you like but it is usually helpful to put $x = 0$ for one value. When $x = 0$, $y = 3 \times 0 + 5$, i.e. $y = 5$. If we let $x = 5$ then $y = 3 \times 5 + 5$, i.e. $y = 20$. We need to plot point (a) $x = 0$, $y = 5$ and point (b) $x = 5$, $y = 20$. This is done in Fig. 3.2. We then join the two points.

It will be noted that in the equation $y = 3x + 5$, the constant term 5 is the point where the line cuts the y axis; the distance between this point and the origin is called the *intercept*. If the constant term is negative, the line cuts the y axis below the origin. The coefficient of x, in this example 3, gives the *gradient* (or *slope*) of the line and indicates that if x increases by one unit, y increases by 3 units. If the coefficient of x is positive (as in this example) the line slopes upwards; if the coefficient of x is negative the line slopes downwards. If we are asked to plot the line $y = 5$, we note that for this line there is no x term. i.e. the coefficient of x is zero, so the line has zero gradient. The line is therefore a horizontal line through the point $y = 5$. This line is drawn in Fig. 3.3.

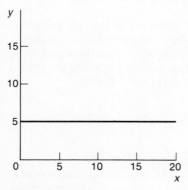

Fig. 3.3

Fig. 3.4

The line $x = 3$ would be a vertical line through the point $x = 3$. This line is plotted in Fig. 3.4.

Exercise 8

Plot the lines: (i) $y = 4x - 2$; (ii) $p = 20 - 2q$; (iii) $y = 4$

NON-LINEAR RELATIONSHIPS

In the example and exercise above we have assumed that a linear relationship existed. A linear relationship is when the graph is a straight line. The equation is of the form $y = mx + c$ – there is no x^2 term. If we have an expression of the form $y = ax^2 + bx + c$, and we plot this, we obtain a curve.

Worked Example 9

Plot the curve $y = x^2 - 9x + 8$.

We need to find y for various values of x.

x	0	1	2	3	4	5	6	7	8	9
x^2	0	1	4	9	16	25	36	49	64	81
$-9x$	0	-9	-18	-27	-36	-45	-54	-63	-72	-81
8	8	8	8	8	8	8	8	8	8	8
$y = x^2 - 9x + 8$	8	0	-6	-10	-12	-12	-10	-6	0	8

This curve is called a *parabola*. The values obtained where the curve cuts the x axis ($x = 1$ and $x = 8$) are the *solutions* of the quadratic equation $x^2 - 9x + 8 = 0$; when the curve $y = x^2 - 9x + 8$ cuts the x axis, $y = 0$.

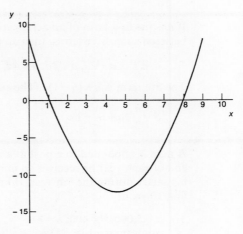

Fig. 3.5

Exercise 9

Plot the curve $R = 12q - q^2$. When is R zero and what is the maximum value of R?
(*Answers:* 0, 12; 36)

SERIES

In business transactions a series of payments may have to be paid. If the payments are all equal it is a simple matter to calculate the total payment, but if the payments increase either by a constant amount or by a constant proportion, the calculation of the total payments is rather more difficult.

(a) Payments increasing by a constant amount

Suppose that the first payment is £1,000 and that the next payment is £100 more than the previous payment, i.e. the payments are: £1,000; £1,100; £1,200; £1,300; and so on. A series where the arithmetic difference between successive terms is constant is called an **Arithmetic Progression**.

(b) Payments increasing by a constant proportion

Suppose that the first payment is again £1,000 and suppose that the next payment is 10% more than the previous payment, i.e. the second payment is £1,000 + 10% of £1,000 = £1,000 + £100 = £1,100. The third payment is £1,100 + 10% of £1,100 = £1,100 + £110 = £1,210. The fourth payment is £1,210 + 10% £1,210 = £1,210 + £121 = £1,331, and so on. The successive payments are therefore £1,000; £1,100; £1,210; £1,331; £1,464.10; and so on. In this series the ratio of a term to the previous term is constant and is equal to 1.1. As a check

$$\frac{1,100}{1,000} = 1.1, \qquad \frac{1,210}{1,100} = 1.1, \qquad \frac{1,331}{1,210} = 1.1.$$

A series where the ratio of successive terms is constant is called a **Geometric Progression**.

19

ARITHMETIC PROGRESSION

If a is the first term of an arithmetic progression and d is the difference between successive terms, then an arithmetic progression is

$$a, a+d, a+2d, a+3d, a+4d, a+5d, \ldots$$

The nth term is $a+(n-1)d$. The sum of n terms may be shown to be

$$S = \frac{n}{2}(2a+(n-1)d).$$

Worked Example 10

A man is appointed to a post at a salary of £5,000 per annum; at the end of each year he receives an annual increment of £400. Calculate: (i) his salary during the 8th year; (ii) the total salary received during the first 10 years.

(i) $a=5,000$, $d=400$, $n=8$
nth term $= a+(n-1)d$
8th term $= 5,000 + (8-1) \times 400 = 5,000 + 7 \times 400 = £7,800$

(ii) $a=5,000$, $d=400$, $n=10$

Sum of n terms $= \dfrac{n}{2}(2a+(n-1)d)$

$$S = \frac{10}{2}(2 \times 5,000 + (10-1) \times 400) = 5(10,000 + 3,600)$$

$$= 5 \times 13,600 = £68,000$$

Exercise 10

A man agrees to repay a debt by annual payments. The first payment is £500, and each year he reduces the payment by £50. Find the total payments over an 8-year period and the payment in the 7th year. (*Answers:* £2,600; £200)

GEOMETRIC PROGRESSION

If a is the first term of a geometric progression and r is the common ratio, then a geometric progression is

$$a, ar, ar^2, ar^3, ar^4, \ldots$$

The nth term is ar^{n-1}. The sum of n terms may be shown to be

$$S = a\frac{(1-r^n)}{(1-r)}$$

This formulae should be used if $r < 1$.
If $r > 1$,

$$S = a\frac{(r^n-1)}{(r-1)}$$

is more suitable.

Worked Example 11

Find the sum of n terms of the series

$$A, A(1+i), A(1+i)^2, A(1+i)^3, \ldots, A(1+i)^{n-1}$$

(this series is important in compound interest problems)

$a = A$, $r = 1 + i$; assuming i is positive then $r > 1$. We use

$$S = a\frac{(r^n - 1)}{(r - 1)}$$

$$S = A\frac{((1+i)^n - 1)}{((1+i) - 1)} = A\frac{((1+i)^n - 1)}{i}$$

Exercise 11

(i) A man makes a series of annual payments starting at £1,000 and increasing at 5% p.a. Find the total payments over a ten-year period.

(ii) A man makes three payments at yearly intervals, an initial payment of £1,000. After 3 years the total of the payments is £4,000. If the payments increase at a constant percentage, find the percentage. (*Answers:* £12,577.89; 30.27%)

LOGARITHMS

The introduction of the calculator has meant that logarithms are little used for multiplication and division. Most calculators have a square-root key: more sophisticated calculators enable you to find cube and other roots. If your calculator does not have x^y or log keys, you will have to use logarithm tables. Logarithms are not now in the syllabus of some 'O'-level mathematics examinations, so you may not have met logarithms before. In these circumstances it may be desirable for you to buy a calculator with these keys.

Use of logarithm tables

The usual logarithm tables are to base 10. A logarithm of a number is the power by which 10 must be raised to produce that number.

$$10^3 = 1,000 \qquad \log 1,000 = 3.0000$$
$$10^2 = 100 \qquad \log 100 = 2.0000$$
$$10^1 = 10 \qquad \log 10 = 1.0000$$
$$10^0 = 1 \qquad \log 1 = 0.0000$$
$$10^{-1} = 0.1 \qquad \log 0.1 = -1.0000 \text{ (usually written } \bar{1}.0000)$$
$$10^{-2} = 0.01 \qquad \log 0.01 = -2.0000 \text{ (usually written } \bar{2}.0000)$$

For numbers other than those above we have to use tables. To find the logarithm of 476, this number lies between 100 and 1,000. From the list above the logarithm lies between 2 and 3; thus log 476 will be 2 plus a quantity, the logarithmic tables give 0.6776 for 476. Thus the logarithm of 476 is 2.6776. The whole number part (2 in this example) is called the *characteristic* and the decimal part the *mantissa*.

Scientific calculators give you logarithms – you simply enter 476 and press the logarithm key and the answer is 2.677607. If you plan to buy such a calculator make certain the logarithms are to base 10.

Use of logarithms to find square roots and other roots

Suppose you have to find a cube root of a number, say the cube root of 476. The method is to find the logarithm of $476 = 2.6776$, divide by $3 = 0.8925$, and then antilog the result, the antilogarithm of 0.8925 is 7.808. If we have to find, say, the square root or other root of a decimal the procedure is slightly more complicated. To find the square root of 0.476, the logarithm is $\bar{1}.6776$ we need to divide this by 2. We note that $\bar{1}.6776$ is $-1 + 0.6776$. To be able to divide by 2 we change this to $-2 + 1.6776$ and on dividing by 2 we obtain $-1 + 0.8388 = \bar{1}.8388$; the antilog of $\bar{1}.8388$ is 0.6899.

Logarithm of a power

A result we need is $\log a^p = p \log a$.
Now $a^p = a \times a \times a \ldots \times a$, so that a appears p times.
We remember that when we use logarithms for multiplication we add the logarithms.
Thus $\log a^p = \log a + \log a + \log a + \log a + \ldots + \log a$
On the right-hand side $\log a$ appears p times.
Hence $\log a^p = p \log a$

Worked Example 12

Find the value of n if $(1.07)^n = 2$

We take logarithms of both sides: $\log (1.07)^n = \log 2$
Using the result above $\log (1.07)^n = n \log 1.07$

$$n \log 1.07 = \log 2 \ \therefore \ n = \frac{\log 2}{\log 1.07}$$

If your calculator has a logarithm key, proceed as follows: Enter 2, press log key, press $\div$ key, enter 1.07, press log key then press $=$ key. You should obtain 10.244768.

If your calculator does not have a logarithm key, you will have to use the logarithm tables. From tables $\log 2 = 0.3010$, $\log 1.07 = 0.0294$.

$$n = \frac{\log 2}{\log 1.07} = \frac{0.3010}{0.0294} = 10.24$$

Exercise 12

Find the value of n if $(1.1)^n = 3$
(*Answer:* 11.52)

Worked Example 13

If $(1 + x)^4 = 1.65$, find x

As $(1 + x)$ is raised to the 4th power, then $(1 + x)$ is equal to the fourth root of 1.65. How to find the fourth root depends upon the sophistication of your calculator.

If your calculator has an $x^{\frac{1}{y}}$ key, enter 1.65, press $x^{\frac{1}{y}}$ key (you may have to press the second function or inverse key first), enter 4, then press $=$ key. You should obtain 1.1333681.

If your calculator has an x^y key, enter 1.65, press x^y key, enter 0.25 ($x^{\frac{1}{4}} = x^{0.25}$), then press $=$ key. You should obtain 1.1333681.

If your calculator has a log key, enter 1.65, press log key, press $\div$ key, enter 4, then press $=$ key, you should obtain 0.0543709, then antilog (to antilog you usually have to press second function or inverse key and then press log key). You should obtain 1.1333681.

If you have to use logarithm tables, look up logarithm of 1.65 = 0.2175, divide by 4 = 0.0544, then antilog = 1.133.

Having found the fourth root of 1.65 by some method, then

$$1 + x = 1.1333681 \qquad x = 0.1334$$

Exercise 13

If $(1 + x)^{10} = 2.3$ find x.
(*Answer:* 0.0869)

MATRIX ALGEBRA

Before studying this section, check whether matrix algebra is included in the syllabus. At the time of writing, matrix algebra is included in the CACA syllabus for numerical analysis and data processing but not in the syllabuses of other major professional bodies at the foundation level.

Matrix algebra is included in many 'O'-level mathematics syllabuses, so this topic should be familiar to some; however the application of matrix algebra to business problems is considered in Chapter 12 and is outside the usual 'O'-level syllabus.

VECTORS

Worked Example 14

A firm makes two products A and B. The products are sold in cartons of ten. Suppose that the firm receives the following orders:
(i)   2 cartons of A and 4 cartons of B
(ii) 5 cartons of A and 9 cartons of B
(iii) 6 cartons of B
(iv) 4 cartons of A

These orders could be recorded as:

[2   4]   [5   9]   [0   6]   [4   0]

The data has been recorded in the form of a vector and as it is in a row it is called a *row vector*. The numbers in the vector are called *elements*.

Addition and subtraction of vectors

If we add the vectors above we obtain the total of the four orders.

[2   4] + [5   9] + [0   6] + [4   0] = [11   19]

If an order [3   6] were returned the net amount to draw from stock would be:

[11   19] − [3   6] = [8   13]

Multiplication of vectors

(i) By a scalar
If we have two orders, each for 2 cartons of product A and 4 cartons of product B, i.e. 2 of [2   4], these orders total

2[2   4] = [4   8]

Each element is multiplied by the scalar quantity 2.

(ii) By another vector
Suppose the price of a carton of product A is £36 and of product B is

£20. Then the total cost of order [2 4] would be $2 \times 36 + 4 \times 20$ $= £152$.

We may represent the prices as a *column vector* $\begin{bmatrix} 36 \\ 20 \end{bmatrix}$

and the total cost of order is given by the vector multiplication

$$[2 \quad 4] \begin{bmatrix} 36 \\ 20 \end{bmatrix}$$

The multiplication is shown on the right

$$[2 \quad 4] \begin{bmatrix} 36 \\ 20 \end{bmatrix} = 2 \times 36 + 4 \times 20 = 152$$

We multiply first element of the row vector with first element of the column vector and *add* the second element of the row vector multiplied by the second element of the column vector.

Exercise 14

Using vector multiplication find the costs of orders [5 9], [0 6] and [4 0] if the costs are £36 and £20 respectively.
(*Answers:* 360; 120; 144)

Equations in vector form

An equation can be written in vector form, e.g.

Worked Example 15

$$[2 \quad 5] \begin{bmatrix} x \\ 3 \end{bmatrix} = 23 \text{ is the same as } 2x + 15 = 23$$

The value of x is 4.

Exercise 15

Solve the equation

$$[3 \quad 7] \begin{bmatrix} x \\ -2 \end{bmatrix} = 13$$

(*Answer:* 9)

MATRICES

Worked Example 16

In making products A and B there are three processes: cutting, shaping and polishing. Suppose that the times in minutes per carton are as follows:

	Cutting	Shaping	Polishing
Product A	8	10	30
Product B	4	8	25

This may be expressed as a matrix of processing times:

$$\begin{bmatrix} 8 & 10 & 30 \\ 4 & 8 & 25 \end{bmatrix}$$

To find the production time for the order [2 4] we have

$2 \times 8 + 4 \times 4 \quad = 32$ minutes of cutting
$2 \times 10 + 4 \times 8 = 52$ minutes of shaping
$2 \times 30 + 4 \times 25 = 160$ minutes of polishing

In matrix form this is written:

$$[2 \quad 4] \begin{bmatrix} 8 & 10 & 30 \\ 4 & 8 & 25 \end{bmatrix} = [32 \quad 52 \quad 160]$$

To carry out the multiplication we multiply the order vector by the first column of the time matrix, we then multiply the order vector by the second column of the time matrix, and so on.

Suppose now that there are four separate orders. We can express the four orders as a matrix:

$$\begin{bmatrix} 2 & 4 \\ 5 & 9 \\ 0 & 6 \\ 4 & 0 \end{bmatrix}$$

We can find the manufacturing times for each order.

$$\begin{bmatrix} 2 & 4 \\ 5 & 9 \\ 0 & 6 \\ 4 & 0 \end{bmatrix} \begin{bmatrix} 8 & 10 & 30 \\ 4 & 8 & 25 \end{bmatrix} = \begin{bmatrix} 32 & 52 & 160 \\ 76 & 122 & 375 \\ 24 & 48 & 150 \\ 32 & 40 & 120 \end{bmatrix}$$

You will see that the first row of the right-hand matrix was found earlier by multiplying the first row of the order matrix [2 4] with the time matrix. The second row of the right-hand matrix is found by multiplying the second row of the order matrix [5 9] with the time matrix, and so on.

Note: For two matrices to be able to be multiplied together the number of columns of the first matrix must be equal to the number of rows of the second matrix.

It is possible to multiply more than two matrices/vectors together. If for example the costs per minute of cutting, shaping and polishing are 14, 20 and 35 pence respectively, then for order [2 4] the manufacturing cost is:

$$[2 \quad 4] \begin{bmatrix} 8 & 10 & 30 \\ 4 & 8 & 25 \end{bmatrix} \begin{bmatrix} 14 \\ 20 \\ 35 \end{bmatrix}$$

To work out this multiplication, we note that we have already multiplied the first row vector and the time matrix. So we have:

$$[32 \quad 52 \quad 160] \begin{bmatrix} 14 \\ 20 \\ 35 \end{bmatrix} = 7{,}088 = \pounds70.88$$

Unit matrix	The matrix

$$I = \begin{bmatrix} 1 & 0 \\ 0 & 1 \end{bmatrix}$$

or

$$I = \begin{bmatrix} 1 & 0 & 0 \\ 0 & 1 & 0 \\ 0 & 0 & 1 \end{bmatrix}$$

is called the *unit matrix*.

A matrix when multiplied by a unit matrix is unchanged, for example:

$$\begin{bmatrix} 2 & 4 \\ 5 & 9 \end{bmatrix}\begin{bmatrix} 1 & 0 \\ 0 & 1 \end{bmatrix} = \begin{bmatrix} 2 & 4 \\ 5 & 9 \end{bmatrix} \qquad \begin{bmatrix} 1 & 0 \\ 0 & 1 \end{bmatrix}\begin{bmatrix} 2 & 4 \\ 5 & 9 \end{bmatrix} = \begin{bmatrix} 2 & 4 \\ 5 & 9 \end{bmatrix}$$

Inverse of a matrix

IF

$$A = \begin{bmatrix} 8 & 3 \\ 2 & 1 \end{bmatrix}$$

then the *inverse* of A, written A^{-1}, is such that

$$AA^{-1} = I \text{ or } A^{-1}A = I,$$

where I is the unit matrix.

At the introductory level candidates are only expected to invert a 2×2 matrix. You need to learn the inverse of the 2×2 matrix.

$$\text{If } A = \begin{bmatrix} a & b \\ c & d \end{bmatrix}$$

then the inverse is

$$\begin{bmatrix} \dfrac{d}{k} & \dfrac{-b}{k} \\ \dfrac{-c}{k} & \dfrac{a}{k} \end{bmatrix}$$

where $k = ad - bc$.

If

$$A = \begin{bmatrix} 8 & 2 \\ 3 & 1 \end{bmatrix}$$

then $k = ad - bc = 8 \times 1 - 2 \times 3 = 2$.

Then

$$A^{-1} = \begin{bmatrix} 0.5 & -1 \\ -1.5 & 4 \end{bmatrix}$$

Check

$$AA^{-1} = \begin{bmatrix} 8 & 2 \\ 3 & 1 \end{bmatrix} \begin{bmatrix} 0.5 & -1 \\ -1.5 & 4 \end{bmatrix} = \begin{bmatrix} 1 & 0 \\ 0 & 1 \end{bmatrix} = I,$$

the unit matrix.

Exercise 16

Find the inverse of

$$\begin{bmatrix} 2 & 1 \\ 6 & 5 \end{bmatrix}$$

(*Answer:* $\begin{bmatrix} 1.25 & -0.25 \\ -1.5 & 0.5 \end{bmatrix}$)

Solution of equations using matrices

The method of solution is shown in Worked Example 17.

Worked Example 17

Solve the equations: $8x + 2y = 26$
$3x + y = 10$

These equations may be written in matrix form:

$$\begin{bmatrix} 8 & 2 \\ 3 & 1 \end{bmatrix} \begin{bmatrix} x \\ y \end{bmatrix} = \begin{bmatrix} 26 \\ 10 \end{bmatrix}$$

The inverse of

$$\begin{bmatrix} 8 & 2 \\ 3 & 1 \end{bmatrix}$$

was found above to be

$$\begin{bmatrix} 0.5 & -1 \\ -1.5 & 4 \end{bmatrix}$$

Premultiply both sides by the inverse

$$\begin{bmatrix} 0.5 & -1 \\ -1.5 & 4 \end{bmatrix}:$$

$$\begin{bmatrix} 0.5 & -1 \\ -1.5 & 4 \end{bmatrix} \begin{bmatrix} 8 & 2 \\ 3 & 1 \end{bmatrix} \begin{bmatrix} x \\ y \end{bmatrix} = \begin{bmatrix} 0.5 & -1 \\ -1.5 & 4 \end{bmatrix} \begin{bmatrix} 26 \\ 10 \end{bmatrix}$$

Now

$$\begin{bmatrix} 0.5 & -1 \\ -1.5 & 4 \end{bmatrix} \begin{bmatrix} 8 & 2 \\ 3 & 1 \end{bmatrix} = \begin{bmatrix} 1 & 0 \\ 0 & 1 \end{bmatrix}$$

$$\begin{bmatrix} 1 & 0 \\ 0 & 1 \end{bmatrix} \begin{bmatrix} x \\ y \end{bmatrix} = \begin{bmatrix} 0.5 & -1 \\ -1.5 & 4 \end{bmatrix} \begin{bmatrix} 26 \\ 10 \end{bmatrix}$$

On multiplying left- and right-hand sides we obtain:

$$\begin{bmatrix} x \\ y \end{bmatrix} = \begin{bmatrix} 3 \\ 1 \end{bmatrix}$$

If two vectors are equal then the elements are equal, i.e.

$$x = 3 \quad y = 1$$

Exercise 17

Using matrices solve: $2x + y = 11$
$$6x + 5y = 35$$

(*Answer:* $x = 5$; $y = 1$)

C. SOLUTIONS TO EXERCISES

S1

(i) $2 \times 7 \times a^3 \times a \times b^4 \times b^3 = 14a^{(3+1)} b^{(4+3)} = 14a^4 b^7$

(ii) $\dfrac{9a^2bc^3}{b^2} \times \dfrac{b^3}{18a^2c^2} = \dfrac{9a^2 b^{(1+3)} c^3}{18a^2 b^2 c^2} = \dfrac{b^2 c}{2}$

S2

(i) $2 \times 12 - 2 \times 10 + -8 \times -24 = 24 - 20 + 192 = 196$

(ii) $(2 - (-8))((-8)^2 - 2 \times (-24) \times 12) = 10 \times (64 + 576) = 6{,}400$

(iii) $\dfrac{-(-8) + \sqrt{(-8)^2 - 4 \times 2 \times (-24)}}{2 \times 2} = \dfrac{8 + \sqrt{64 + 192}}{4} = \dfrac{8 + \sqrt{256}}{4}$

$$= \dfrac{8 + 16}{4} = \dfrac{24}{4} = 6$$

S3

Rise $= 150 \times \dfrac{7.5}{100} = 11.25$, new earnings $150 + 11.25 = £161.25$

$$A \times 1.075 = 172, \quad A = \dfrac{172}{1.075} = £160$$

S4

20% of $9{,}000 = 9{,}000 \times \dfrac{20}{100} = 1{,}800$. Amount left $= 9{,}000 - 1{,}800$

$= 7{,}200$. Divide in ratio 15:11:10, 36 parts, each part $= 200$.
A receives $15 \times 200 = £3{,}000$; B $11 \times 200 = £2{,}200$; C $10 \times 200 = £2{,}000$

S5

(i) $3x + 5 = 14 \quad \therefore 3x = 14 - 5 \quad \therefore 3x = 9 \quad \therefore x = 3$

(ii) $3 - 4x = 6x - 7 \quad \therefore 3 + 7 = 6x + 4x \quad \therefore 10 = 10x \quad \therefore x = 1$

(iii) $5 - 3(x - 4) = 2(3 - 2x) \quad \therefore 5 - 3x + 12 = 6 - 4x$
$\quad \therefore 4x - 3x = 6 - 12 - 5 \quad \therefore x = -11$

(iv) To clear fractions multiply by 30: $6(2x + 3) - 15(x - 2) - 20 = 4x$
$\quad \therefore 12x + 18 - 15x + 30 - 20 = 4x$
$\quad \therefore 18 + 30 - 20 = 4x + 15x - 12x \quad \therefore 28 = 7x \quad \therefore x = 4$

S6

  (i) Add, $7x = 35$ $\therefore x = 5$, $2 \times 5 + 3y = 22$ $\therefore 3y = 22 - 10$ $\therefore y = 4$

  (ii) Multiply second equation by 4, $4x + 4y = 24$ then add to first equation, $7x = 28$ $\therefore x = 4$, $4 + y = 6$ $\therefore y = 6 - 4$ $\therefore y = 2$

  (iii) Multiply first equation by 2, second equation by 3, $12x - 10y = 90$; $12x - 21y = 123$, subtract second from first, $-10y - (-21y) = 90 - 123$ $\therefore -10y + 21y = -33$ $\therefore 11y = -33$ $\therefore y = -3$
$6x - 5(-3) = 45$ $\therefore 6x + 15 = 45$ $\therefore 6x = 30$ $\therefore x = 5$

S7

  (i) Product $P = +24$, sum $S = -11$; factors -3, -8,
$(x - 3)(x - 8) = 0$, $x = 3$ $x = 8$

  (ii) $P = -45$, $S = -4$; -9, $+5$, $(x - 9)(x + 5) = 0$, $x = 9$ $x = -5$

  (iii) Use formulae $a = 3$, $b = 5$, $c = -4$, $x = \dfrac{-5 \pm \sqrt{5^2 - 4 \times 3 \times (-4)}}{2 \times 3}$

$$x = \frac{-5 \pm \sqrt{25 + 48}}{6} = \frac{-5 \pm \sqrt{73}}{6} = \frac{-5 \pm 8.544}{6},$$

$$x = 0.591 \quad x = -2.257$$

S8

  (i) $x = 0$, $y = -2$; $x = 10$, $y = 38$; plot and join points – Fig. 3S.1
  (ii) $q = 0$, $p = 20$; $q = 8$, $p = 4$; plot and join points – Fig. 3S.2
  (iii) $y = 4$ is a horizontal line which cuts y axis at 4 – Fig. 3S.3

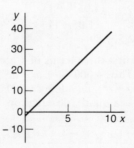
Fig. 3S.1

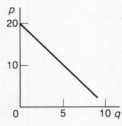
Fig. 3S.2

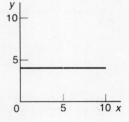
Fig. 3S.3

S9

q	0	2	4	6	8	10	12
$12q$	0	24	48	72	96	120	144
$-q^2$	0	-4	-16	-36	-64	-100	-144
$R = 12q - q^2$	0	20	32	36	32	20	0

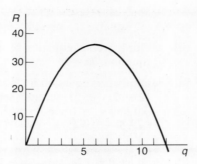

Fig. 3S.4

The curve cuts the horizontal axis when $q = 0$ and $q = 12$. From Fig. 3S.4 we can see that maximum occurs when $q = 6$ and the maximum value of R is 36.

S10

$a = 500, d = -50, n = 8, S = \dfrac{8}{2}(2 \times 500 + (8 - 1) \times (-50)) = 2,600$

7th term $= 500 + (7 - 1) \times (-50) = 200$.

S11

(i)   Payments $1,000, 1,050, 1,102.50, \ldots, a = 1,000, r = \dfrac{1,050}{1,000} = 1.05$

$S = 1,000\dfrac{(1.05^{10} - 1)}{(1.05 - 1)} = 1,000\dfrac{(1.6288946 - 1)}{0.05} = \dfrac{628.8946}{0.05}$
$= 12,577.89$

(ii) The first three terms of a geometric progression are a, ar and ar^2. Thus $4,000 = a + ar + ar^2$; now $a = 1,000$, so $4,000 = 1,000 (1 + r + r^2)$. Divide by $1,000$ $\therefore 4 = 1 + r + r^2$ $\therefore r^2 + r - 3 = 0$ this is a quadratic equation; solving

$r = \dfrac{-1 \pm \sqrt{1^2 - 4 \times 1 \times (-3)}}{2} = \dfrac{-1 \pm \sqrt{13}}{2} = \dfrac{-1 \pm 3.60555}{2}$

$r = 1.3027$ (we ignore the negative solution), $1,000$ increases to $1,302.7$, i.e. a 30.27% increase. (**Note:** compound interest is covered in Chapter 12.)

S12

$n \log 1.1 = \log 3, n = \dfrac{\log 3}{\log 1.1} = \dfrac{0.4771}{0.0414} = 11.52$

S13

$1 + x =$ tenth root of 2.3 – with a sophisticated calculator this is easily found to be 1.0868579. With tables look up log of $2.3 = 0.3617$, divide by $10 = 0.0362$, antilog $= 1.087$. $x = 0.0869$ or 0.087.

S14

$$[5 \quad 9] \begin{bmatrix} 36 \\ 20 \end{bmatrix} = 5 \times 36 + 9 \times 20 = 360$$

$$[0 \quad 6] \begin{bmatrix} 36 \\ 20 \end{bmatrix} = 0 \times 36 + 6 \times 20 = 120$$

$$[4 \quad 0] \begin{bmatrix} 36 \\ 20 \end{bmatrix} = 4 \times 36 + 0 \times 20 = 144$$

S15

$$[3 \quad 7] \begin{bmatrix} x \\ -2 \end{bmatrix} = 13 \quad \therefore 3x - 14 = 13 \quad \therefore 3x = 14 + 13 \quad \therefore x = 9$$

S16

$$A = \begin{bmatrix} 2 & 1 \\ 6 & 5 \end{bmatrix} \qquad k = 2 \times 5 - 1 \times 6 = 4$$

$$A^{-1} = \begin{bmatrix} \dfrac{5}{4} & \dfrac{-1}{4} \\[2mm] \dfrac{-6}{4} & \dfrac{2}{4} \end{bmatrix} = \begin{bmatrix} 1.25 & -0.25 \\ -1.5 & 0.5 \end{bmatrix}$$

S17

$$\begin{aligned} 2x + y &= 11 \\ 6x + 5y &= 35 \end{aligned} \qquad \begin{bmatrix} 2 & 1 \\ 6 & 5 \end{bmatrix} \begin{bmatrix} x \\ y \end{bmatrix} = \begin{bmatrix} 11 \\ 35 \end{bmatrix}$$

The inverse of $\begin{bmatrix} 2 & 1 \\ 6 & 5 \end{bmatrix}$ was found in Exercise 16 = $\begin{bmatrix} 1.25 & -0.25 \\ -1.5 & 0.5 \end{bmatrix}$

$$\begin{bmatrix} 1.25 & -0.25 \\ -1.5 & 0.5 \end{bmatrix} \begin{bmatrix} 2 & 1 \\ 6 & 5 \end{bmatrix} \begin{bmatrix} x \\ y \end{bmatrix} = \begin{bmatrix} 1.25 & -0.25 \\ -1.5 & 0.5 \end{bmatrix} \begin{bmatrix} 11 \\ 35 \end{bmatrix}$$

$$\begin{bmatrix} 1 & 0 \\ 0 & 1 \end{bmatrix} \begin{bmatrix} x \\ y \end{bmatrix} = \begin{bmatrix} 5 \\ 1 \end{bmatrix} \qquad \begin{bmatrix} x \\ y \end{bmatrix} = \begin{bmatrix} 5 \\ 1 \end{bmatrix} \qquad \begin{aligned} x &= 5 \\ y &= 1 \end{aligned}$$

A STEP FURTHER

A useful additional source for those requiring a more mathematical approach is Tennant-Smith, *Mathematics for the Manager*. The chapters of this book combine the revision of mathematics with the application of mathematics to business problems. To revise various mathematical techniques using this book the index should be used.

Chapter 4

Presentation of data

A. GETTING STARTED

Most examination syllabuses expect candidates to be able to present data diagrammatically using either graphs or charts. In most 'O'-level mathematics courses *bar charts* and *pie charts* are taught, thus some material of this chapter will be familiar to you. Different methods of presentation are covered in the various chapters. *Cumulative frequency curves (ogives)* in Chapter 5; *scatter diagrams* in Chapter 6; *histograms* in Chapter 6; *Venn diagrams* in Chapter 9; *break-even charts* in Chapter 12.

B. ESSENTIAL PRINCIPLES

FORMATION OF A FREQUENCY TABLE

Worked Example 1

A machine produces the following number of rejects in each successive period of five minutes:

16	21	26	24	11	24	17	25	26	13
27	24	26	3	27	23	24	15	22	22
12	22	29	21	18	22	28	25	7	17
22	28	19	23	23	22	3	19	13	31
23	28	24	9	20	33	30	23	20	8

Construct a *frequency distribution* from these data, using seven class intervals of equal width.

ICMA (part) May 1982

If the question states 'seven class intervals of equal width', you must do as instructed otherwise marks are lost.

The first step is to find the smallest and largest values ie 3 and 33. The range is $33 - 3 = 30$, divide by $7 = 4.71$, so the group width should be 5. The boundaries of the groups should be unambiguous, e.g. 0 and under 5, 5 and under 10, etc.; it is then clear where 5 would be entered. Boundaries such as 0 to 5, 5 to 10 would be penalized by an examiner.

The best method of finding the frequencies, is to use a tally as in Table 4.1.

Table 4.1

Number of rejects	Tally	Frequency
0 and under 5	||	2
5 and under 10	|||	3
10 and under 15	||||	4
15 and under 20	JHT ||	7
20 and under 25	JHT JHT JHT JHT	20
25 and under 30	JHT JHT |	11
30 and under 35	|||	3
Total		50

Exercise 1

Your company manufactures components for use in the production of motor vehicles. The number of components produced each day over a forty-day period is illustrated in Table 4.2.

Table 4.2

553	526	521	528	538
523	538	546	524	544
532	554	517	549	512
528	523	510	555	545
524	512	525	543	532
533	519	521	536	534
541	535	531	551	535
519	530	549	518	531

Group the data into five classes.

(CACA [ACCA]-part, Dec. 1983)

HISTOGRAM

Worked Example 2

Draw a histogram of the data of Worked Example 1.

The data of Worked Example 1 can be represented diagrammatically as in Fig. 4.1, with the area of the rectangles being proportional to the frequency. This is achieved by making the height of the rectangles equal to the frequency. Remember to label the axes and to give a title to the histogram. The source of the data should be given. The diagram should be neatly drawn using a ruler and pencil.

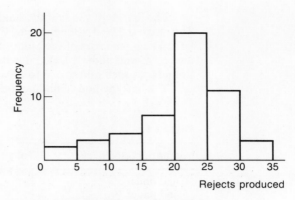

Fig. 4.1   Rejects produced by machine

Exercise 2   Draw a histogram of the data of Exercise 1.

In Worked Example 1 the class intervals were of *equal width*. If the class intervals are *unequal* an adjustment is necessary before the histogram can be drawn. Worked Example 3 explains the method then appropriate.

Worked Example 3   **Table 4.3**

Annual earnings of employees of a company

Earnings (£1,000)	Number of employees
2.5 and under 5	25
5 and under 7.5	42
7.5 and under 10	52
10 and under 15	78
15 and under 20	18

In Fig. 4.2 the data is drawn with the heights of the rectangles proportional to the numbers in each class interval. This form of a 'histogram' is incorrect and is heavily penalized by examiners. The *unequal* class intervals must be adjusted. The group '10 and under 15' is twice the width of the earlier groups. The method is to divide this group into two groups; 10 and under 12.5, 12.5 and under 15, and to assume that the 78 employees are equally divided (39) between the two groups. The 15 and under 20 group is dealt with in the same way. Figure 4.3 is the correct histogram.

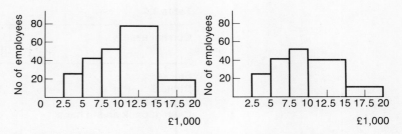

Fig. 4.2   *Annual earnings*   Fig. 4.3   *Annual earnings*

Exercise 3

Draw a histogram for the data in Table 4.4 which relates to the length of life of bad debts.

Table 4.4

Number of working days	Number of debts
1 and under 5	30
5 and under 10	20
10 and under 20	32
20 and under 30	14
30 and over	4

FREQUENCY POLYGON

If we join the mid-points of the top of the rectangles of the histogram in Fig. 4.1 we obtain a *frequency polygon*. This is shown in Fig. 4.4.

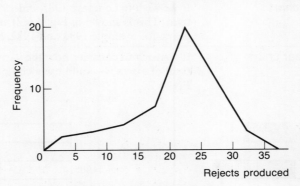

Fig. 4.4   *Rejects produced by machine*

CHARTS

Suppose we were asked to present diagrammatically the information given in Table 4.5, taken from the accounts of Marks and Spencer plc.

Table 4.5

Current assets	1984 (£m)	1985 (£m)
Stocks	157.5	185.2
Debtors	92.1	100.0
Investments	57.9	102.0
Cash at bank and in hand	10.5	16.6
Total	318.0	403.8

1. Simple bar chart

A simple bar chart which gives the current assets for 1985 is shown in Fig. 4.5.

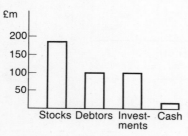

Source: Company accounts

Fig. 4.5   Marks & Spencer plc current assets 1985
Source: company accounts

£m   Key: ■ 1984   □ 1985

Source: Company accounts

Fig. 4.6   Marks & Spencer plc current assets 1984 & 1985
Source: company accounts

2. Multiple bar chart

If we wish to compare 1985 and 1984, a multiple bar chart could be used. This is shown in Fig. 4.6. It would be possible to compare more years, for example 1983 and 1982, by adding to the multiple bar chart.

3. Component bar chart

If we wish to consider how the various types of assets make up the total of assets we could use a component bar chart. This is shown in Fig. 4.7.

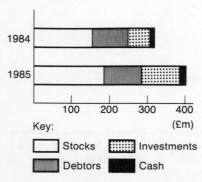

Key:
Stocks Investments
Debtors Cash

Fig. 4.7   Marks & Spencer plc current assets

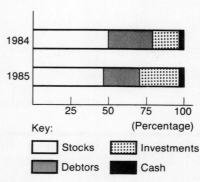

Key:
Stocks Investments
Debtors Cash

Fig. 4.8   Marks & Spencer plc current assets

4. Percentage component bar chart

The *component bar chart* shows how the total of the assets are split between the various types of assets. The *percentage component bar chart*, by expressing each asset as a percentage of the total, shows the change in the *relative* importance of each asset. However, the disadvantage is that it does *not* show the changes in the *absolute total* which are available from the component bar chart. A percentage component bar chart is shown in Fig. 4.8.

To construct a percentage component bar chart it is necessary to find the percentages; for 1984 the percentage for stocks is

$$\frac{157.5}{318.0} \times 100 = 49.5\%.$$

The remaining percentages are shown in Table 4.6.

Table 4.6

Current assets	1984 (£m.)	1985 (£m.)	1984 (%)	1985 (%)	1984 (angle)	1985 (angle)
Stocks	157.5	185.2	49.5	45.9	178	165
Debtors	92.1	100.0	29.0	24.8	104	89
Investments	57.9	102.0	18.2	25.3	66	91
Cash at bank and in hand	10.5	16.6	3.3	4.1	12	15
Total	318.0	403.8	100.0	100.1*	360	360

*rounding error

5. Pie chart

A pie chart is an alternative presentation to a percentage component bar chart, and is usually preferred. In an examination room you may have to choose between the two. If so, only choose the pie chart if you

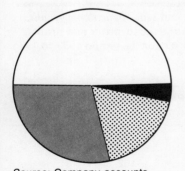

Source: Company accounts

Key:

☐ Stocks   ▦ Investments

▨ Debtors   ■ Cash

Fig. 4.9 Marks & Spencer plc current assets 1984
Source: company accounts

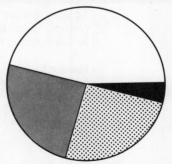

Source: Company accounts

Key:

☐ Stocks   ▦ Investments

▨ Debtors   ■ Cash

Fig. 4.10 Marks & Spencer plc current assets 1985
Source: company accounts

have a pair of compasses and a protractor; a freehand pie chart will be penalized (it is no good saying you left your protractor at home – examiners are hard-hearted!). To construct a pie chart it is necessary to work out the *angles*; for stocks for 1984, the angle is

$$\frac{157.5}{318.0} \times 360 = 178.$$

The remaining angles are shown in the Table 4.6. Pie charts for 1984 and 1985 are shown in Figs. 4.9 and 4.10.

Exercise 4

Use an appropriate method to represent the data in Table 4.7 diagrammatically.

Table 4.7

Marks & Spencer plc:
Shareholders owning more than 100,000 shares

Type of owner	Number of shareholders
Insurance companies	94
Banks and nominee companies	231
Pension funds	122
Individuals	189
Others	146
Total	782

Source: Company accounts

(ICSA (part) June 1984)

GRAPHS

To illustrate the use of *ordinary graphs* and *semi-logarithmic graphs*, consider the data in Table 4.8 which shows the turnover and profits before taxation for Marks and Spencer plc for the period 1977 to 1985.

Table 4.8

Year	Turnover (£m.)	Profits before taxation (£m.)
1977	1,065	102
1978	1,254	118
1979	1,473	162
1980	1,668	174
1981	1,873	181
1982	2,209	222
1983	2,517	239
1984	2,868	279
1985	3,213	303

Source: Company accounts

We can plot this data on ordinary graph paper. You should have years on the horizontal axis, turnover and profit on the vertical axis.

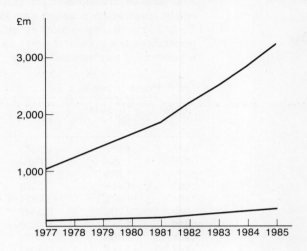

Fig. 4.11   Marks & Spencer plc turnover & profit before taxation
Source: company accounts

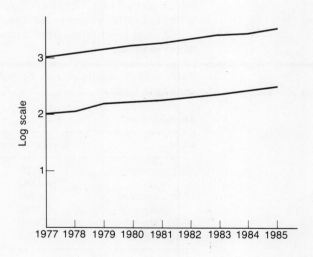

Fig. 4.12   Marks & Spencer plc turnover & profit before taxation
Source: company accounts

Ordinary graph paper shows the *absolute* changes. It is not easy to tell from Fig. 4.11 whether profits are rising at the same rate of growth as turnover. To compare *rates of change* a *semi-logarithmic* graph should be used. Special semi-logarithmic graph paper is available but it is unlikely to be supplied in an examination set by a professional body – the professional bodies find it almost impossible to ensure that all centres receive copies of this graph paper. A semi-logarithmic graph can be constructed using ordinary graph paper by taking logarithms. To find logarithms you must use either the tables supplied or preferably use a calculator which has a log key. Table 4.9 gives logarithms of turnover and profit.

Table 4.9

Year	Turnover £m.	log	Profits before taxation £m.	log
1977	1,065	3.0273	102	2.0086
1978	1,254	3.0983	118	2.0719
1979	1,473	3.1682	162	2.2095
1980	1,668	3.2222	174	2.2405
1981	1,873	3.2725	181	2.2577
1982	2,209	3.3442	222	2.3464
1983	2,517	3.4009	239	2.3784
1984	2,868	3.4576	279	2.4456
1985	3,213	3.5069	303	2.4814

Years should be on the horizontal axis, the logarithms should be on the vertical axis. The graph is plotted in Fig. 4.12. The rates of growth of turnover and profits can now be compared.

A semi-logarithmic graph cannot be used if the data has zero or negative values.

Exercise 5

Table 4.10 shows the average of daily telegraphic transfer per £1 sterling in London.

Table 4.10

Date	Swiss francs	Deutschemark
1977: January	4.2701	4.102
July	4.1543	3.934
1978: January	3.8398	4.094
July	3.4143	3.892
1979: January	3.3479	3.708
July	3.7205	4.122

(i)   Using the standard graph paper provided, plot the value of £1 in Swiss and German currencies on a semi-logarithmic scale.
(ii) Interpret the graph.
(iii) Comment on the merits and demerits of a semi-logarithmic graph compared with a natural scale graph with particular reference to the data given.

(ICMA Nov. 1980)

Z-CHART

This is a graph which has three components: (i) monthly data; (ii) cumulative data; and (iii) a moving annual total. The basic data is given in the first two columns of Table 4.11.

Table 4.11

Sales

Month	1984	1985	Cumulative total for 1985	Moving annual total
Jan.	78	84	84	621
Feb.	67	79	163	633
Mar.	59	72	235	646
Apr.	46	57	292	657
May	35	45	337	667
Jun.	29	34	371	672
Jul.	27	33	404	678
Aug.	23	28	432	683
Sept.	43	52	484	692
Oct.	57	68	552	703
Nov.	71	77	629	709
Dec.	80	85	714	714

The *cumulative total* for 1985 is obtained by adding the second data column cumulatively, e.g. $84 + 79 = 163$, $163 + 72 = 235$, and so on. To obtain the *moving annual total* for January add the months from February 1984 to January 1985 inclusive $= 621$. The moving annual total for February 1985 is found by *adding* to the January 1985 total of 621 the figure for February 1985 and *subtracting* the figure for

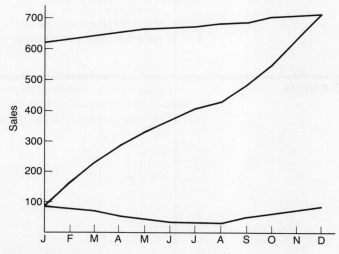

Fig. 4.13   Z–chart

February 1984, i.e. $621 + 79 - 67 = 633$. Proceed in this way. As a check on accuracy the cumulative total and moving annual total for December – 714 – must be the same.

The Z-chart is plotted in Fig. 4.13. The bottom line represents the monthly data for 1985, and shows any seasonality in the data: the sales are lower in the summer months. The middle line is the cumulative total and records the cumulative sales for the year. The moving annual total, the top line, shows that sales are increasing.

Exercise 6

The manufacturing costs of Prodco Ltd for the period January 1980 to December 1981 are displayed in Table 4.12.

Table 4.12

Month		Manufacturing costs (£000)	Month		Manufacturing costs (£000)
1980	Jan.	35.8	1981	Jan.	43.9
	Feb.	33.6		Feb.	40.1
	Mar.	35.5		Mar.	46.0
	Apr.	37.5		Apr.	48.7
	May	37.2		May	48.9
	Jun.	34.6		Jun.	46.0
	Jul.	36.3		Jul.	48.3
	Aug.	36.0		Aug.	47.9
	Sep.	35.4		Sep.	48.2
	Oct.	30.4		Oct.	47.0
	Nov.	36.1		Nov.	49.4
	Dec.	37.5		Dec.	51.6

(i) Draw a Z-chart (sometimes known as a Zee chart) using the above data.
(ii) What is the purpose of such a chart? Use the diagram that you have just drawn to illustrate your answer.

(CACA (ACCA) (part) June 1982)

LORENZ CURVE

A *Lorenz curve* is used to show *inequality* diagrammatically. It is often used to show inequality of income and wealth but can be employed in other areas. Consider the data given in Table 4.13.

Table 4.13

Annual income of employees of a company

Range of income (£)	Number of employees	Total income (£000)
2,000 and under 4,000	141	502
4,000 and under 6,000	680	3,372
6,000 and under 8,000	763	5,344
8,000 and under 10,000	464	4,058
10,000 and under 15,000	210	2,425
15,000 and under 20,000	150	2,537
20,000 and under 30,000	20	480
30,000 and over	7	298

The first step is to obtain *cumulative totals* for both sets of data. This is shown in columns (ii) and (v) in Table 4.14. The next step is to express each of the cumulative entries as a *percentage* of the total. This is shown in columns (iii) and (vi). For example, to obtain 5.8% for column (iii) the calculation is

$$\frac{141}{2,435} \times 100.$$

To obtain the next figure of 33.7%, the calculation is

$$\frac{821}{2,435} \times 100,$$

and so on.

Table 4.14

	Employees			Income	
Number	Cumulative numbers	Cumulative percentage	Income	Cumulative income	Cumulative percentage
(i)	(ii)	(iii)	(iv)	(v)	(vi)
141	141	5.8	502	502	2.6
680	821	33.7	3,372	3,874	20.3
763	1,584	65.1	5,344	9,218	48.5
464	2,048	84.1	4,058	13,276	69.8
210	2,258	92.7	2,425	15,701	82.6
150	2,408	98.9	2,537	18,238	95.9
20	2,428	99.7	480	18,718	98.4
7	2,435	100.0	298	19,016	100.0

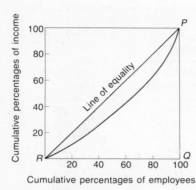

Fig. 4.14   *Lorenz curve*

Columns (iii) and (vi) are plotted in Fig. 4.14. Column (iii) is plotted on the horizontal axis, column (vi) on the vertical axis. The diagonal line drawn from the origin is called 'the line of equality'. If 10% of the employees earned 10% of the income, 20% of the employees earned 20% of the income and so on, then *all* the data would lie on the line of perfect equality. The plot shows that the actual curve is away from this line, indicating inequality: the further the curve is away from the diagonal line, i.e. the greater the shaded area, the greater the inequality. (In Exercise 7 you will find that wealth data shows greater inequality than income data.) There is a measure of inequality called the *Gini coefficient*. This is the ratio of the shaded area to the area of the triangle PQR. You are not normally expected to find the Gini coefficient in foundation level courses.

Exercise 7

The figures given in Table 4.15 relate to the distribution of identified personal wealth in Great Britain and were taken from the *Annual Abstract of Statistics*.

Table 4.15

Ranges of net wealth (£)	Number of cases (thousands)	£ thousand million
under 1,000	3,410	2.0
1,000 and under 3,000	4,775	8.6
3,000 and under 5,000	2,223	8.7
5,000 and under 10,000	4,131	30.3
10,000 and under 15,000	2,166	26.5
15,000 and under 20,000	757	13.3
20,000 and under 25,000	415	9.6
25,000 and under 50,000	640	21.7
50,000 and under 100,000	229	15.4
100,000 and under 200,000	65	9.2
200,000 and over	26	11.8
Total	18,837	157.1

Present the information in the form of a Lorenz curve and comment on the results.

(ICMA (part) Nov. 1979)

C. SOLUTIONS TO EXERCISES

S1, S2

Smallest number 510, largest 555. Use classes of equal width as this makes drawing the histogram or other work easier. There is no unique solution.

510 and under 520	7
520 and under 530	10
530 and under 540	12
540 and under 550	7
550 and under 560	4
Total	40

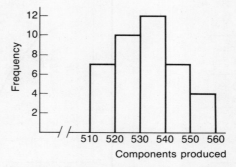

Fig. 4S.1   Daily production of components

S3, S4

3. See Fig. 4S.2. 4. Pie chart (Fig. 4S.3), followed by a component bar chart, are the best choices.

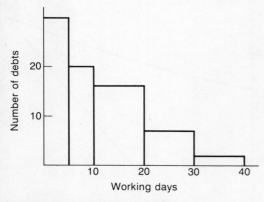

Fig. 4S.2   Length of life of bad debts

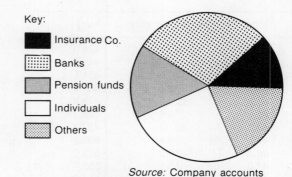

Key:
- ■ Insurance Co.
- ▦ Banks
- ▨ Pension funds
- □ Individuals
- ▩ Others

Source: Company accounts

Fig. 4S.3   Marks & Spencer plc large shareholders
Source: company accounts

Swiss francs	Deutsche marks
0.6304	0.6130
0.6185	0.5948
0.5843	0.6121
0.5333	0.5902
0.5248	0.5691
0.5706	0.6151

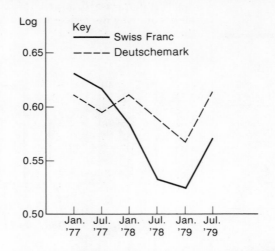

Fig. 4S.4

(ii) Apart from period July 1977 to Jan. 1978 both currencies have moved at roughly the same rate of change.

(iii) See Outline Answer 3 below.

1981	Monthly	Cum. monthly	MAT
Jan.	43.9	43.9	434.0
Feb.	40.1	84.0	440.5
Mar.	46.0	130.0	451.0
Apr.	48.7	178.7	462.2
May	48.9	227.6	473.9
Jun.	46.0	273.6	485.3
Jul.	48.3	321.9	497.3
Aug.	47.9	369.8	509.2
Sep.	48.2	418.0	522.0
Oct.	47.0	465.0	538.6
Nov.	49.4	514.4	551.9
Dec.	51.6	566.0	566.0

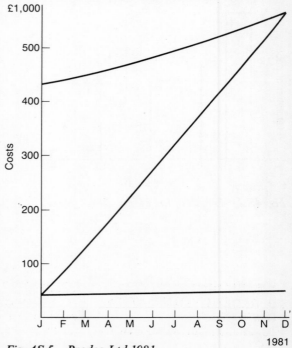

Fig. 4S.5   Prodco Ltd 1981

Cases			Wealth		
No.	Cu. no.	Cu. %	Wealth	Cu. wealth	Cu. %
3,410	3,410	18.1	2.0	2.0	1.3
4,775	8,185	43.5	8.6	10.6	6.7
2,223	10,408	55.3	8.7	19.3	12.3
4,131	14,539	77.2	30.3	49.6	31.6
2,166	16,705	88.7	26.5	76.1	48.4
757	17,462	92.7	13.3	89.4	56.9
415	17,877	94.9	9.6	99.0	63.0
640	18,517	98.3	21.7	120.7	76.8
229	18,746	99.5	15.4	136.1	86.6
65	18,811	99.9	9.2	145.3	92.5
26	18,837	100.0	11.8	157.1	100.0

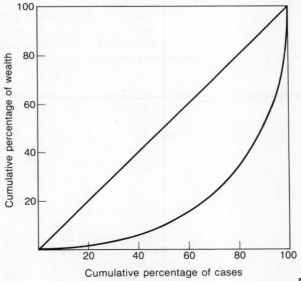

Source: Annual Abstract

Fig. 4S.6

Wealth is very unevenly distributed.

D. RECENT EXAMINATION QUESTIONS

Q1

A survey was made of cars parked in car parks of a town. For each car, a record was made of whether it was British made (B) or made by a Foreign manufacturer (F). Its age was also recorded in years. Thus B1 stands for a car up to 1 year old made by a British manufacturer.

The results of surveying 120 vehicles were as follows:

B6	B9	F8	B5	F9	B12	B5	B6	F2	F3	B6	F4
B10	B1	B6	B1	F3	B6	F6	F8	B4	B4	B12	F9
F1	B6	F5	B3	B8	F2	B4	B11	F5	B1	B6	B7
B1	B8	B4	B5	B10	F4	B5	B7	B6	F5	B8	F6
B2	F6	F1	F9	B7	B5	F6	B3	B1	B6	F1	F12
B2	B9	B2	B2	F7	B2	F5	B2	B3	B10	F2	B3
F10	F2	F3	F5	F1	B3	B9	F1	B2	B7	B2	F1
F11	F4	B12	F3	B1	B8	B12	F11	F2	B8	B5	F10
B1	F2	F7	B9	B3	B4	F3	F2	B5	B11	B2	B7
B5	B10	B8	F7	F1	B10	F2	B5	B11	F5	B5	F4

(a) Make two tables, one for British and one for foreign, showing the number of cars of each age.

(b) Draw histograms to show the ages of British and foreign cars, using 0 up to 1, 1 up to 2, ..., 11 up to 12 years as intervals.

(c) It is required to summarize the results of this survey in a table which labels all cars aged 6 and upwards 'older' and all cars aged 5 or less 'newer'. British and foreign cars to be kept separate in this table.

Design a suitable two-way table, providing appropriate space for totals and headings. Enter the summary figures in your two-way table. Make a brief comment on the figures which you think important.

(RSA (adapted) June 1984)

Q2

Table Q2 shows stoppages of work due to industrial disputes.

Table Q2

Duration in man-days	Number of stoppages beginning in 1981	Aggregate number of working days lost in these stoppages (to nearest thousand)
Under 250	593	56
250 and under 500	166	61
500 and under 1,000	170	123
1,000 and under 5,000	304	683
5,000 and under 25,000	80	779
25,000 and under 50,000	12	425
50,000 and over	13	2,116

Source: Employment Gazette

Illustrate the above data by means of a Lorenz curve.

(RSA (part) June 1984)

Q3 (a) Plot the following data on arithmetic scale graph paper.

Period	1	2	3	4	5	6	7	8
Data	200	320	640	1,180	2,080	4,050	6,480	9,030

(b) Explain how to draw a semi-logarithmic curve when no semi-logarithmic graph paper is available. Plot the above data as a semi-logarithmic curve on arithmetic scale graph paper. State the advantages and disadvantages of using semi-log scale over the more usual arithmetic scale.

(ICMA (adapted) Nov. 1974)

Q4 Your company is in the course of preparing its published accounts and the Chairman has requested that the assets of the Company be prepared in a component bar chart for the last five years. The data for this task is contained in the Table Q4.

Table Q4

Asset	1978 (£'000s)	1979 (£'000s)	1980 (£'000s)	1981 (£'000s)	1982 (£'000s)
Property	59	59	65	70	74
Plant and machinery	176	179	195	210	200
Stock and work-in-progress	409	409	448	516	479
Debtors	330	313	384	374	479
Cash	7	60	29	74	74

(i) Construct the necessary component bar charts.

(ii) Comment upon the movements in the assets over the five-year period.

[This is the question actually set, but for practice you could draw all the other charts.]

(CACA (ACCA) (part) June 1984)

Q5 Table Q5 shows the value of the sales (in £'000s) of an engineering company for 1982 and 1983.

Table Q5

	1982	1983		1982	1983
January	20.5	23.2	July	39.4	42.7
February	21.6	27.9	August	38.6	44.9
March	29.4	33.1	September	35.2	39.3
April	33.8	41.5	October	30.1	36.2
May	30.3	32.3	November	29.4	24.9
June	38.9	36.4	December	25.3	29.6

49

(i) Construct a Z-chart to display this data.
(ii) Explain what the Z-chart shows about this data.
(iii) Describe briefly the construction and use of one other chart or graph which could be used to display this data.

(LCCI, Dec. 1985)

E. OUTLINE ANSWERS TO EXAM QUESTIONS

A1

Age	1	2	3	4	5	6	7	8	9	10	11	12	Total
British	7	9	6	5	10	9	5	6	4	5	3	4	73
Foreign	7	8	5	4	6	4	3	2	3	2	2	1	47

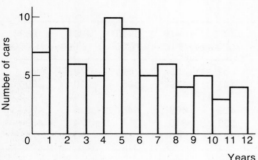

Fig. 4S.7   *Age of British cars*

Fig. 4S.8   *Age of foreign cars*

	Newer cars (5 years and under)	Older cars (6 years or more)	Totals
British	37	36	73
Foreign	30	17	47
Totals	67	53	120

More British cars but a higher proportion of newer foreign cars.

A2

Stoppages			Working days lost		
No.	Cu. no.	Cu. %	No.	Cu. no.	Cu. %
593	593	44.3	56	56	1.3
166	759	56.7	61	117	2.8
170	929	69.4	123	240	5.7
304	1,233	92.2	683	923	21.8
80	1,313	98.1	779	1,702	40.1
12	1,325	99.0	425	2,127	50.1
13	1,338	100.0	2,116	4,243	100.0

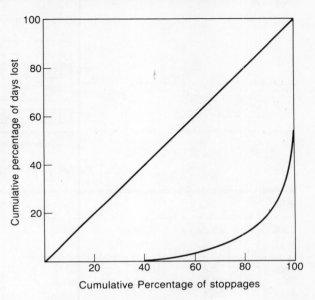

Source: Employment Gazette

Fig. 4S.9   Source: Employment Gazette

A3

Data	200	320	640	1,180	2,080	4,050	6,480	9,030
Log	2.3010	2.5051	2.8062	3.0719	3.3181	3.6075	3.8116	3.9557

Advantages (i) Shows rate of change.   (ii) Very useful if range of data is large.   (iii) If graph straight line, shows constant rate of growth; if upward curve, shows increasing rate of growth; if falling curve, shows decreasing rate of growth.

Disadvantages (i) Not as well known as arithmetic scale.   (ii) Zero and negative values cannot be shown.   (iii) If special semi-log graph paper not available, have to know how to use logarithms.

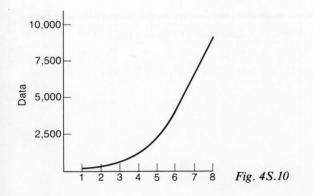

Fig. 4S.10

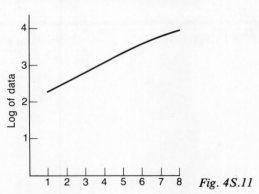

Fig. 4S.11

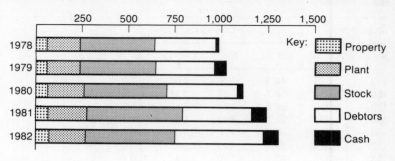

Fig. 4S.12

A5

1988	Monthly	Cum. monthly	MAT
Jan.	23.2	23.2	375.2
Feb.	27.9	51.1	381.5
Mar.	33.1	84.2	385.2
Apr.	41.5	125.7	392.9
May	32.3	158.0	394.9
Jun.	36.4	194.4	392.4
Jul.	42.7	237.1	395.7
Aug.	44.9	282.0	402.0
Sep.	39.3	321.3	406.1
Oct.	36.2	357.5	412.2
Nov.	24.9	382.4	407.7
Dec.	29.6	412.0	412.0

Fig. 4S.13

(i) As the moving average curve is upward sloping, sales increased in 1983. Sales are higher in the summer months.

(ii) Plot both series on the same graph.

A STEP FURTHER   Mulholland and Jones, *Fundamentals of Statistics*, Ch. 2.

Chapter 5

Measures of location and dispersion

MEASURES OF LOCATION

In the analysis of data we are often asked to find an 'average'. In statistics there are three main averages in use:

(i) the mode;   (ii) the median;   (iii) the mean

The *mode* is the item that occurs *most frequently*. For example, in the case of households with children, the most frequently occurring number of children in a family in the United Kingdom is 2. We say that for households with children the mode number of children is 2. The mode, however, is not used very much in statistical work and is not often set in examinations. The method for finding the mode is shown in the answer to Exam Question 5.

The *median* is found by arranging the data *in order of magnitude*. The median is then the *middle* item. For example, suppose the sales commissions (£) of 15 representatives were as follows:

23, 16, 31, 77, 21, 14, 32, 6, 155, 9, 36, 24, 5, 27, 19

Placing the data in order of magnitude, we have:

5, 6, 9, 14, 16, 19, 21, 23, 24, 27, 31, 32, 36, 77, 155

The middle item is the eighth value – 23 (there are 7 values smaller than 23 and 7 values larger than 23). The median value is therefore £23.

The *mean* is the statistical name for what is commonly called *the average*. The mean of the sales commission data is:

$$= \frac{23 + 16 + 31 + 14 + 21 + 77 + 32 + 6 + 155 + 9 + 36 + 24 + 5 + 27 + 19}{15}$$

$$\text{ie mean} = \frac{495}{15} = 33$$

From this analysis it will be noticed that the *mean* sales commission is £33, whereas the *median* sales commission is only £23.

The mean is higher because two values 77 and 155 were much higher than the rest. In other words, the data is *skewed* (see p. 64). In practice the mean exceeds the median for most data on incomes, profits and wealth. Trade unions, in negotiating for higher wages, prefer to use the *median* income, arguing that the median income more accurately represents the 'typical' employee than the mean income.

The mode, median and the mean are called *measures of location*. In addition to finding the 'average', it is also useful to know how widely spread the data is around the average. For the sales commission data the mean income was £33 but the highest and lowest values were £155 and £5 respectively. Trade unions, when looking at the pay of their members, are clearly anxious that their average pay does not fall behind those in comparable occupations. However, they are also interested in the gap between the low paid and the high paid.

MEASURES OF DISPERSION

The two main measures of 'spread' used in statistics are the quartile deviation and the standard deviation. These are called *measures of dispersion*.

Quartiles and quartile deviation

The *lower quartile* and the *upper quartile*, together with the *median*, divide a distribution into four equal parts. In the data on sales commissions the *lower quartile*, denoted by Q_1, is a quarter of the way along the distribution, in this case the fourth value – 14; the *upper quartile*, denoted by Q_3, is three-quarters of the way along the distribution, in this case the twelfth value – 32.

The *quartile deviation* is defined as the average difference between the lower quartile and the median, and the upper quartile and the median. The formula for the quartile deviation is

$$\text{Quartile deviation} = \frac{Q_3 - Q_1}{2}$$

The quartile deviation for the sales commission data is:

$$\text{Quartile deviation} = \frac{32 - 14}{2} = \frac{18}{2} = 9$$

Another measure of dispersion is the *interquartile range*. This is equal to $Q_3 - Q_1$, and is twice the quartile deviation.

B. ESSENTIAL PRINCIPLES

Most data in examination questions is presented in the form of a frequency table. You are then asked to calculate named measures of location and dispersion. The data in the following Worked Example (Table 5.1) is in the form of a frequency table.

Worked Example 1

Table 5.1

£	Number of orders
Value of orders	
under 20	12
20 and under 40	26
40 and under 60	60
60 and under 80	36
80 and under 100	25
100 and under 150	22
150 and under 200	13
200 and under 300	4
300 and over	2
Total	200

Obtain: (i) median; (ii) lower quartile; (iii) upper quartile; (iv) quartile deviation; (v) lowest decile; (vi) highest decile.

There are two methods of obtaining these measures: (a) a graphical method; (b) a calculation method. Examination questions usually specify which method you have to use. It is necessary, therefore, to learn **both** methods. The calculation method is explained in Worked Example 2.

Whichever method you follow, you must first place the data into a cumulative frequency table (Table 5.2).

Table 5.2

Less than (£)	Cumulative frequency
20	12
40	38
60	98
80	134
100	159
150	181
200	194
300	198
400	200

The figure of 38 for 'under 40', is obtained by adding the 26 from the group '20 and under 40' to the 12 from the group 'under 20'. This process is repeated. As a check on accuracy, the cumulative total for the last group – 200 – must be the same as the total from the frequency table.

GRAPHICAL METHOD OF FINDING MEDIAN, QUARTILES AND DECILES

The next step is to plot the data as in Fig. 5.1. We plot 12 against £20, 38 against £40, 98 against £60, and so on. The resulting graph is a *cumulative frequency curve* (sometimes called an *ogive*). You should always use graph paper. The cumulative data must be plotted on the vertical axis. You should label both axes, give the graph a title, and when appropriate give the source. Marks are lost for poor graphs and no labels.

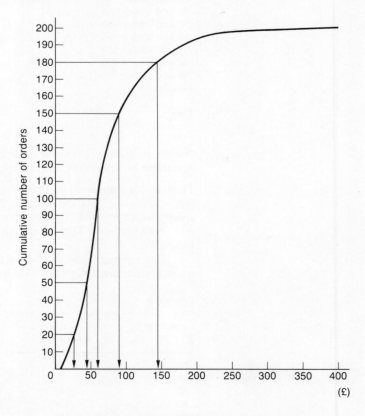

Fig. 5.1   Cumulative distribution of value of orders

The *median* is the middle item, and is at the position $n/2$, where n is the total number of observations (in some textbooks $n+1$ is used instead of n. This is incorrect when the data is in percentage form; it is therefore suggested that you use n rather than $n+1$ in all cases).

The *lower quartile* is at position $n/4$, the *upper quartile* is at position $\dfrac{3 \times n}{4}$.

The *deciles* divide the distribution into 10 equal parts. The *lowest decile* is at position

$$\frac{n}{10}$$

and the *highest decile* is at position

$$\frac{9 \times n}{10}.$$

In this example $n = 200$. The position of median is 100, lower quartile is at 50, upper quartile is at 150, lowest decile is at 20 and the highest decile is 180.

The next step is to draw a horizontal line from the points 50, 100, 150 on the vertical axis, until you meet the curve, then draw vertical lines down to the horizontal axis, as shown on Fig. 5.1. The values at which these lines cut the horizontal axis give the values of the lower quartile (about £45), the median (about £60) and the upper quartile (about £90) respectively.

Similarly we can find the positions of the lowest decile and the highest decile by drawing horizontal lines at 20 and 180 respectively. The vertical lines give value of the lowest decile (about £30) and the highest decile (about £145).

$$\text{The quartile deviation} = \frac{Q_3 - Q_1}{2} = \frac{90 - 45}{2} = £22.5$$

Exercise 1

A property dealer classified the properties which he had sold in a six-monthly period, by value, in Table 5.2.

Table 5.2

Value of property (£'000)	Number of properties
10 and less than 15	2
15 and less than 20	6
20 and less than 25	14
25 and less than 30	21
30 and less than 35	33
35 and less than 40	19
40 and less than 45	5

(a) Draw an ogive for the above data.
(b) Use the ogive to find: (i) median; (ii) the quartiles; (iii) quartile deviation; (iv) lowest and highest deciles.
(*Answer in £'000:* 31; 26 and 35; 4.5; 20.5 and 38)

CALCULATION OF MEDIAN, QUARTILES AND DECILES

Using the data of Worked Example 1 calculate: (i) median; (ii) the quartiles; (iii) quartile deviation; (iv) highest decile.

We need the cumulative frequency table (Table 5.3) again.

Table 5.3

Less than (£)	Cumulative frequency	
20	12	
40	38	
		← Q_1
60	98	
		← M
80	134	
		← Q_3
100	159	
		← D_9
150	181	
200	194	
300	198	
400	200	

As we found on p. 57 the position of the median is 100; that of the lower quartile is 50; the upper quartile 150; and the highest decile 180. The next step is to insert arrows in Table 5.3. The median is at position 100, therefore insert an arrow marked M between 98 and 134. The median occurs in the class interval £60 to £80, which is termed the median class interval. Similar arrows are placed for the upper and lower quartiles (Q_3 and Q_1) and the highest decile (D_9).

The formula for the median is:

$$\text{Median} = L_1 + (L_2 - L_1)\frac{(n/2 - \Sigma f_1)}{(\Sigma f_2 - \Sigma f_1)}$$

(Σ is the mathematical term which means 'sum')

where L_1 is the lower boundary of the median class interval,

     L_2 is the upper boundary of the median class interval,

     n is the total number of observations,

     Σf_1 is the cumulative frequency corresponding to L_1,

     Σf_2 is the cumulative frequency corresponding to L_2.

To find these values refer to the arrow marked M in Table 5.3.

L_1 and Σf_1 are above the arrow, $L_1 = 60$, $\Sigma f_1 = 98$.

L_2 and Σf_2 are below the arrow, $L_2 = 80$, $\Sigma f_2 = 134$.

$n = 200$

$$M = \text{median} = 60 + (80 - 60)\frac{(200/2 - 98)}{(134 - 98)}$$

$$M = 60 + \frac{20 \times 2}{36} \quad \therefore \ M = 60 + 1.11 \quad \therefore \ M = £61.11$$

The formula for each quartile is very similar to that for the median.

$$Q_1 = L_1 + (L_2 - L_1)\frac{(n/4 - \Sigma f_1)}{(\Sigma f_2 - \Sigma f_1)}$$

where L_1, L_2, Σf_1, Σf_2, relate to the quartiles rather than the median. Above the arrow marked Q_1, $L_1 = 40$, $\Sigma f_1 = 38$. Below arrow, $L_2 = 60$, $\Sigma f_2 = 98$.

$$Q_1 = 40 + (60 - 40)\frac{(200/4 - 38)}{(98 - 38)}$$

$$Q_1 = 40 + \frac{20 \times 12}{60} \quad \therefore Q_1 = 40 + 4 \quad \therefore Q_1 = £44$$

Similarly,

$$Q_3 = L_1 + (L_2 - L_1)\frac{(3n/4 - \Sigma f_1)}{(\Sigma f_2 - \Sigma f_1)}$$

Above the arrow marked Q_3, $L_1 = 80$; $\Sigma f_1 = 134$. Below arrow, $L_2 = 100$, $\Sigma f_2 = 159$

$$Q_3 = 80 + (100 - 80)\frac{(150 - 134)}{(159 - 134)}$$

$$Q_3 = 80 + \frac{20 \times 16}{25} \quad \therefore Q_3 = 80 + 12.8 \quad \therefore Q_3 = £92.80$$

$$\text{Quartile deviation} = \frac{Q_3 - Q_1}{2} = \frac{92.80 - 44}{2} = £24.40$$

The formula for the deciles are of the same form:

$$D_9 = L_1 + (L_2 - L_1)\frac{(9n/10 - \Sigma f_1)}{(\Sigma f_2 - \Sigma f_1)}$$

Above the arrow marked D_9, $L_1 = 100$, $\Sigma f_1 = 159$. Below arrow, $L_2 = 150$, $\Sigma f_2 = 181$

$$D_9 = 100 + (150 - 100)\frac{(180 - 159)}{(181 - 159)}$$

$$D_9 100 + \frac{50 \times 21}{22} \quad \therefore D_9 = 100 + 47.73 \quad \therefore D_9 = £147.73$$

Exercise 2

Using the data of Exercise 1 calculate: (i) median; (ii) the quartiles; (iii) quartile deviation; (iv) lowest and highest deciles. (*Answers in £'000:* 31.06; 34.85 and 25.71; 4.57; 20.71 and 38.68)

CALCULATION OF MEAN and STANDARD DEVIATION

Some text books use an assumed mean and a class interval in calculating the mean and the standard deviation. Now that calculators are allowed in almost all examinations it is **not** recommended that this procedure is followed. Candidates obtain a higher success rate using the procedures outlined below.

Calculation of mean

Value of orders (£)	Number of orders
under 20	12
20 and under 40	26
40 and under 60	60
60 and under 80	36
80 and under 100	25
100 and under 150	22
150 and under 200	13
200 and under 300	4
300 and over	2

The mean value ($\bar{x}$) is equal to the total value of all the orders divided by the number of orders. The formula for finding the *mean* is:

$$\bar{x} = \frac{\Sigma fx}{\Sigma f}$$

Where f is the frequency for each class interval, and x is the central value (or mid-point) of each class interval.

The easiest way to find x, the central value, is to add the upper and lower boundaries of a range and divide by 2; for example, in the case of the range '20 and under 40', the central value is

$$\frac{20+40}{2} = 30$$

We are in effect assuming that the average value for orders in the class interval £20 to £40 is £30. Care has to be exercised in choosing the central value for open-ended groups, i.e. the first group 'under 20' and the last group '300 and over'. The value chosen is usually arbitrary, but must be realistic.

The calculation is set out in Table 5.4.

Table 5.4

x	f	fx
15	12	180
30	26	780
50	60	3,000
70	36	2,520
90	25	2,250
125	22	2,750
175	13	2,275
250	4	1,000
350	2	700
Totals $\Sigma f = 200$		$\Sigma fx = 15,455$

The total value of all orders is $\Sigma fx = £15,455$, the total number of orders is $\Sigma f = 200$. Using the formula:

$$\text{Mean} = \bar{x} = \frac{\Sigma fx}{\Sigma f} = \frac{15,455}{200} = 77.275 = £77.28 \text{ (to nearest penny)}$$

Always check to see if your answer is reasonable. If you made a mistake when using your calculator and obtained (say) £772.75, this is far larger than the value of the highest order. You should then recheck your working.

Exercise 3

Using the data of Exercise 1 calculate the mean value of property. (*Answer in £'000:* 30.2)

Calculation of standard deviation

The *standard deviation*, denoted by σ, is found by taking the difference between each observation and the mean, squaring the difference, adding the squares, averaging the squares and finally taking the square root of this average. In mathematical terms this is equal to

$$\sigma = \sqrt{\frac{\Sigma(x - \bar{x})^2}{n}} \qquad \text{for ungrouped data}$$

$$= \sqrt{\frac{\Sigma f(x - \bar{x})^2}{\Sigma f}} \qquad \text{for grouped data}$$

For grouped data the following formula should be used:

$$\sigma = \sqrt{\frac{\Sigma fx^2}{\Sigma f} - \left(\frac{\Sigma fx}{\Sigma f}\right)^2}$$

The standard deviation indicates the spread of a distribution.

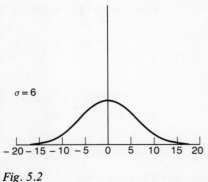

$\sigma = 6$

$\sigma = 2$

Fig. 5.2

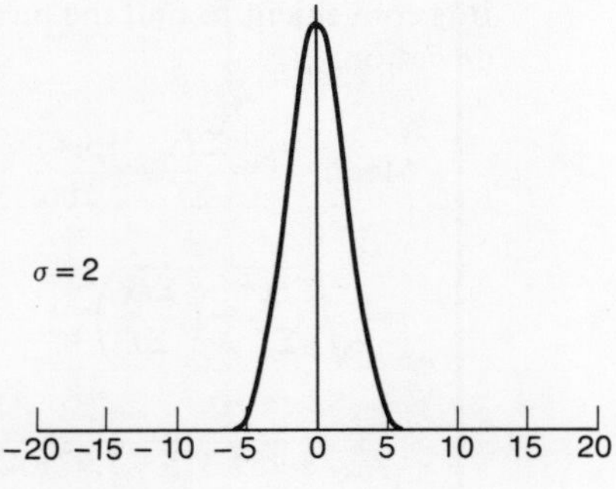

Fig. 5.3

Figures 5.2 and 5.3 show the spread when $\sigma = 6$ and $\sigma = 2$, respectively.

It will be noticed that with $\sigma = 6$ the spread is three times the spread when $\sigma = 2$.

Most examination questions on the calculation of the standard

61

deviation involve grouped data. The method of calculation is explained in Worked Example 4. How to find the standard deviation for ungrouped data is shown in the answer to Exam Question 6 below.

Worked Example 4

Using the data of Worked Example 3, calculate the standard deviation.

To find the standard deviation we need to calculate Σfx^2, Σfx and Σf. In Worked Example 3 we found Σfx *and* Σf (Table 5.4); we now need to find Σfx^2; this means we square x then multiply by f and then add up all the fx^2. A common error is to find fx and then square this and add up the squares i.e. find $\Sigma(fx)^2$. This is a *method* error as opposed to an *arithmetic* error and is usually very heavily penalized by examiners. The tabulation of Worked Example 3 needs to be extended to include a fourth column (Table 5.5).

Table 5.5

x	f	fx	fx²
15	12	180	2,700
30	26	780	23,400
50	60	3,000	150,000
70	36	2,520	176,400
90	25	2,250	202,500
125	22	2,750	343,750
175	13	2,275	398,125
250	4	1,000	250,000
350	2	700	245,000
Totals $\Sigma f = 200$		$\Sigma fx = 15,455$	$\Sigma fx^2 = 1,791,875$

It is convenient to find the mean at the same time as the standard deviation.

$$\text{Mean} = \bar{x} = \frac{\Sigma fx}{\Sigma f} = \frac{15,455}{200} = 77.275 = \text{\pounds}77.28$$

$$\sigma = \sqrt{\frac{\Sigma fx^2}{\Sigma f} - \left(\frac{\Sigma fx}{\Sigma f}\right)^2} \qquad \sigma = \sqrt{\frac{1,791,875}{200} - \left(\frac{15,455}{200}\right)^2}$$

$$= \sqrt{8,959.375 - 5,971.426} \qquad = \sqrt{2,987.949} = \text{\pounds}54.66$$

As a rough check of accuracy, six times the standard deviation should be approximately equal to the range (i.e. difference between largest and smallest values) of the distribution. The range of the central values is £15 to £350 = £335. Now $6 \times 54.66 = 328$. This is not exactly equal to 335, but indicates that no glaring arithmetic error has been made.

If you think that you have made an error, first check to see if you have found $\Sigma(fx)^2$ rather than Σfx^2. Check that you have the first two columns correctly labelled – the central values must be labelled x, the frequencies f. Check the totals, the formula (if you are given a formulae list in the examination room check you have copied this particular formula correctly), the substitution into the formula and the actual arithmetic. If you find that you have a negative quantity for which to find a square root, then you have made a mistake somewhere – you cannot take the square root of a negative quantity.

Exercise 4	Using the data of Exercise 1 calculate the standard deviation. (*Answer in £'000:* 6.724)

CALCULATOR OVERFLOW

The examples and the exercises have been chosen so that students with inexpensive calculators do not have 'overflow' problems, i.e. numbers too large for the calculator to handle. Suppose you are given data on annual salaries where the ranges may include, for example, £10,000 to £12,000, then the central value would be £11,000. To find a standard deviation we have to square this – 121,000,000 – and then multiply by f. However, many calculators cannot handle more than 8 digits. If you have data of this kind (see Exam Question 3 below) then the easiest method is to work in thousands of pounds, i.e. take the range to be 10 to 12. Calculate the mean and standard deviation as in Worked Example 4, and then multiply the final results by 1,000 to obtain the correct answers. You should attempt Exam Question 4 to check that you can deal with large numbers.

OTHER MEASURES OF LOCATION AND DISPERSION

Mean and median are the main measures of location in practice and from the point of view of examination questions. The mode is occasionally examined (see Exam Question 5); other measures are the *geometric mean* and *harmonic mean* which are hardly ever examined.

The *geometric mean* of 12, 18, 16 and 6 is obtained by multiplying the numbers together – $12 \times 18 \times 16 \times 6 = 20,736$ – and then taking the fourth root = 12. The geometric mean is 12. The *Financial Times* index of share prices uses a geometric mean.

The *harmonic mean* involves taking reciprocals. The harmonic mean of 12, 18, 16 and 6 is obtained by adding the reciprocals

$$\frac{1}{12} + \frac{1}{18} + \frac{1}{16} + \frac{1}{6} = \frac{53}{144},$$

take the average, i.e. divide by 4

$$= \frac{53}{576}$$

and take the reciprocal

$$= \frac{576}{53} = 10.868.$$

Thus the harmonic means is 10.87.

$$Quartile\ deviation = \frac{Q_3 - Q_1}{2}$$

(sometimes called the *semi-interquartile range*), together with the *interquartile range* $Q_3 - Q_1$, and the *standard deviation* are the measures of dispersion most frequently examined. Other measures are the *range* and *mean deviation*; these are examined occasionally (see Exam Question 6).

The *range* is simply the largest item less the smallest item. If the sample is 12, 16, 18 and 6, the range is $18 - 6 = 12$. The range is used in quality control (a topic not covered in this book).

The *mean deviation* is the mean of the numerical differences between the observations and the mean. If the sample is 12, 16, 18 and 6, the mean is 13. The numerical differences of the observations from the mean are 1, 3, 5 and 7 (numerical difference ignores sign). The mean of 1, 3, 5 and 7 is 4. Therefore the mean deviation is 4. The mean deviation is not suitable for mathematical treatment and is seldom used.

SKEWNESS

In Fig. 5.4 the data is symmetrically distributed. Only in this case do the mean, median and mode coincide.

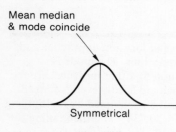

Fig. 5.4

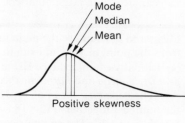

Fig. 5.5

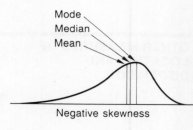

Fig. 5.6

Other data, such as income data, is asymmetrically distributed. Data which is asymmetric is called *skewed*. In skewed data the mean, median and mode do not coincide. When the longer tail points to the right, the data is said to be *positively skewed*. Figure 5.5 is an example of a positively skewed distribution. When the longer tail points to the left, the data is said to be *negatively skewed*. Figure 5.6 is an example of a negatively skewed distribution.

Measures of skewness

There are three measures of skewness; the formulae are given below:

(i) $\dfrac{3\ (\text{mean} - \text{median})}{\sigma}$

(ii) $\dfrac{\text{mean} - \text{mode}}{\sigma}$

where σ is the standard deviation.

(iii) $\dfrac{Q_3 + Q_1 - 2 \times \text{median}}{Q_3 - Q_1}$

where Q_3 and Q_1 are lower and upper quartiles respectively.

In examination questions, finding a measure of skewness (the word *coefficient of skewness* is often used) is usually a final part of a question. If, in the first part of the question, you are asked to find mean and standard deviation, then use either formula (i) or (ii); (i) is easier to use since most students are more proficient at finding the median. On the other hand, if the first part of the question asks for medians and quartiles, then use the quartile measure of skewness given in formula (iii).

When you have found a measure of skewness, the sign of the measure indicates whether the skewness is positive or negative. You should state in your answer whether the data is positively or negatively skewed.

Measures of skewness are independent of the units involved. In the case of income it is possible to compare the skewness of income distributions of different countries even though the currencies differ. It is also possible to compare the skewness of income distributions over a period of time. The first major survey of incomes in the United Kingdom was in 1886. Is the distribution of income as skew today as it was in 1886?

| **Worked Example 5** | Using the data of Worked Example 1, find two measures of skewness. |

(a) Formula (i)

In Worked Examples 1, 2 and 4 we found that the median was £66.11, the mean £77.28 and the standard deviation £54.66. Substitute these values in the formula below:

$$\text{Skewness} = \frac{3\,(\text{mean} - \text{median})}{\sigma} = \frac{3(77.28 - 61.11)}{54.66} = 0.887$$

The distribution is positively skewed.

(b) Formula (ii)

In Worked Example 2 we found $Q_3 = £92.80$, $Q_1 = £44$ and median £61.11. Substituting these values in the formula below:

$$\text{Skewness} = \frac{Q_3 + Q_1 - 2 \times \text{median}}{Q_3 - Q_1}$$

$$= \frac{92.80 + 44 - 2 \times 61.11}{92.80 - 44} = 0.299$$

The distribution is positively skewed.

It will be noted that the two formulae give different values for the coefficient of skewness. Since each is based on different measures, a discrepancy is to be expected.

Exercise 5

Using the data of Exercise 1 find two measures of skewness. (*Answers:* -0.384; -0.171)

COEFFICIENT OF VARIATION

The standard deviation measures dispersion. If we wish to compare the relative dispersion of income distributions of different countries, or of the same country at different times, the standard deviation used alone is of limited use. For instance, in 1886 the mean income of manual workers was about £1.35 per week, but today the mean income is more than a hundred times this figure. Clearly the standard deviation of income today is very much greater than the standard deviation of income in 1886. To compare the relative dispersion of income today with that of 1886 we use the *coefficient of variation*. The formula for the coefficient of variation is shown below:

$$\text{Coefficient of variation} = \frac{\text{standard deviation}}{\text{mean}} = \frac{\sigma}{\bar{x}}$$

Usually this coefficient is expressed as a percentage:

$$\text{Coefficient of variation} = \frac{\sigma \times 100}{\bar{x}}$$

The coefficient of variation is independent of the units. If we compared the relative dispersion of salaries paid in the United Kingdom and France, the different currencies would not matter.

Worked Example 6

Using the data of Worked Example 1, find the coefficient of variation.

In Worked Example 4 we found the mean and standard deviation were £77.28 and £54.66 respectively. Substituting in the formula for coefficient of variation we find:

$$\text{Coefficient of variation} = \frac{\sigma \times 100}{\bar{x}} = \frac{54.66 \times 100}{77.28} = 70.7$$

Exercise 6

Using the data of Exercise 1 find the coefficient of variation. (*Answer:* 22.3)

Less than	Cumulative frequency
15	2
20	8
25	22
30	43
35	76
40	95
45	100

$$\frac{n}{2} = \frac{100}{2} = 50 \qquad \frac{n}{4} = \frac{100}{4} = 25$$

$$\frac{3 \times n}{4} = \frac{3 \times 100}{4} = 75 \qquad \frac{n}{10} = \frac{100}{10} = 10$$

$$\frac{9 \times n}{10} = \frac{9 \times 100}{10} = 90$$

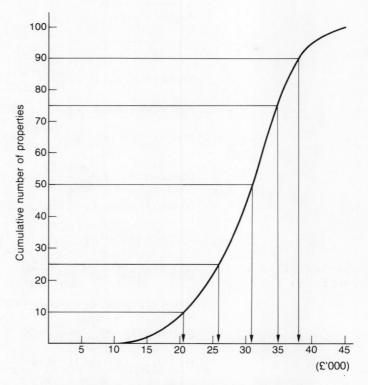

Fig. 5S.1   Cumulative distribution of value of properties

The cumulative frequency curve is shown in Fig. 5S.1. From the curve:
Median $= 31$, $Q_1 = 26$, $Q_3 = 35$

$$\frac{Q_3 - Q_1}{2} = \frac{35 - 26}{2} = \frac{9}{2} = 4.5$$

$D_1 = 20.5$, $D_9 = 38$

Less than	Cumulative frequency	
15	2	
20	8	
		← D₁
25	22	
		← Q₁
30	43	
		← M
		← Q₃
35	76	
		← D₉
40	95	
45	100	

$$\frac{n}{2} = \frac{100}{2} = 50 \qquad \frac{n}{4} = \frac{100}{4} = 25$$

$$\frac{3 \times n}{4} = \frac{3 \times 100}{4} = 75 \qquad \frac{n}{10} = \frac{100}{10} = 10$$

$$\frac{9 \times n}{10} = \frac{9 \times 100}{10} = 90$$

$$\text{Median} = L_1 + (L_2 - L_1)\frac{(n/2 - \Sigma f_1)}{(\Sigma f_2 - \Sigma f_1)}$$

Above M arrow, $L_1 = 30$, $\Sigma f_1 = 43$. Below arrow, $L_2 = 35$, $\Sigma f_2 = 76$

$$\text{Median} = 30 + (35 - 30)\frac{(50 - 43)}{(76 - 43)} = 30 + \frac{5 \times 7}{33} = 30 + 1.06 = 31.06$$

$$Q_1 = L_1 + (L_2 - L_1)\frac{(n/4 - \Sigma f_1)}{(\Sigma f_2 - \Sigma f_1)}$$

Above Q_1 arrow, $L_1 = 25$, $\Sigma f_1 = 22$. Below arrow, $L_2 = 30$, $\Sigma f_2 = 43$

$$Q_1 = 25 + (30 - 25)\frac{(25 - 22)}{(43 - 22)} = 25 + \frac{5 \times 3}{21} = 25 + 0.71 = 25.71$$

$$Q_3 = L_1 + (L_2 - L_1)\frac{(3 \times n/4 - \Sigma f_1)}{(\Sigma f_2 - \Sigma f_1)}$$

Above Q_3 arrow, $L_1 = 30$, $\Sigma f_1 = 43$. Below arrow, $L_2 = 35$, $\Sigma f_2 = 76$

$$Q_3 = 30 + (35 - 30)\frac{(75 - 43)}{(76 - 43)} = 30 + \frac{5 \times 32}{33} = 30 + 4.85 = 34.85$$

$$\frac{Q_3 - Q_1}{2} = \frac{34.85 - 25.71}{2} = \frac{9.14}{2} = 4.57$$

$$D_1 = L_1 + (L_2 - L_1)\frac{(n/10 - \Sigma f_1)}{(\Sigma f_2 - \Sigma f_1)}$$

Above D_1 arrow, $L_1 = 20$, $\Sigma f_1 = 8$. Below arrow, $L_2 = 25$, $\Sigma f_2 = 22$

$$D_1 = 20 + (25 - 20)\frac{(10 - 8)}{(22 - 8)} = 20 + \frac{5 \times 2}{14} = 20 + 0.71 = 20.71$$

$$D_9 = L_1 + (L_2 - L_1)\frac{(9 \times 10/n - \Sigma f_1)}{(\Sigma f_2 - \Sigma f_1)}$$

Above D_9 arrow, $L_1 = 35$, $\Sigma f_1 = 76$. Below arrow, $L_2 = 40$, $\Sigma f_2 = 95$

$$D_9 = 35 + (40 - 35)\frac{(90 - 76)}{(95 - 76)} = 35 + \frac{5 \times 14}{19} = 35 + 3.68 = 38.68$$

S3

Please note only the first three columns are needed for this question.

x	f	fx	fx²
12.5	2	25	312.5
17.5	6	105	1,837.5
22.5	14	315	7,087.5
27.5	21	577.5	15,881.25
32.5	33	1,072.5	34,856.25
37.5	19	712.5	26,718.75
42.5	5	212.5	9,031.25
Totals	100	3,020	95,725

The mean is given by:

$$\bar{x} = \frac{\Sigma fx}{\Sigma f} = \frac{3,020}{100} = 30.2$$

S4

The table is shown in solution S3 above.

$$\sigma = \sqrt{\frac{\Sigma fx_2}{\Sigma f} - \left(\frac{\Sigma fx}{\Sigma f}\right)^2} = \sqrt{\frac{95,725}{100} - \left(\frac{3,020}{100}\right)^2} = \sqrt{957.25 - 912.04}$$

$$= \sqrt{45.21} = 6.724$$

S5

(i) $\text{Skewness} = \dfrac{3(\text{mean} - \text{median})}{\sigma} = \dfrac{3(30.2 - 31.06)}{6.724} = -0.384$

(ii) $\text{Skewness} = \dfrac{Q_3 + Q_1 - 2 \times \text{median}}{Q_3 - Q_1} = \dfrac{34.85 + 25.71 - 2 \times 31.06}{34.85 - 25.71}$

$\qquad = -0.171$

S6

$\text{Coefficient of variation} = \dfrac{\sigma \times 100}{\bar{x}} = \dfrac{6.724 \times 100}{30.2} = 22.3$

D. RECENT EXAMINATION QUESTIONS

In questions involving mean and standard deviation when there are open-ended class intervals, there are a number of correct answers. In the questions that follow, the central values for open-ended classes used in the solutions are given at the end of the question. If you use these central values you should obtain the answers given in the solutions.

Q1

The 1982 profit levels of 50 companies in the electronics industry are shown in the Table Q1.

Table Q1

Profit (£m)	Number of companies
−2 and under 0	4
0 and under 2	11
2 and under 4	17
4 and under 6	12
6 and under 8	4
8 and under 10	2

(i)   Construct a cumulative frequency graph for this data.
(ii) Find the values of the median and quartile deviation.
(iii) What do these values measure?

(LCCI May 1985)

(*Answers in £m:* 3.2; 1.6)

Q2

The following data are supplied (Table Q2):

Table Q2

Weekly earnings (£)	Number of employees
40 and < 50	2
50 and < 60	15
60 and < 70	21
70 and < 80	30
80 and < 90	20
90 and <100	9
100 and <110	3

(a) Calculate: (i) mean; (ii) standard deviation; (iii) coefficient of variation; (iv) median.
(b) Define and explain the use of: (i) standard deviation; (ii) coefficient of variation.

(ICMA (part) May 1981)

(*Answers:* £74; £13.45; 18.2; £74)

A finance company has produced a cumulative frequency distribution (Table Q3) of size of loans made to a randomly selected sample of 300 borrowers.

Table Q3

Size of loan (£)	Number of borrowers
under 1,000	39
under 2,000	139
under 3,000	217
under 4,000	259
under 5,000	285
under 6,000	300

(a) Redraft the data in the form of a grouped frequency distribution.
(b) Calculate the mean and standard deviation for the sample.
   (Central value first group £500)

(ABE (part) June 1982)

(*Answers:* £2,370; £1,330)

Table Q4 shows age at marriage in 1979 (first marriage).

Table Q4

Age	Males (thousands)	Females (thousands)
under 20	24	77
20 and under 25	142	142
25 and under 30	71	36
30 and under 35	23	11
35 and under 40	6	3
40 and under 45	5	3
45 and over	3	2

Source: Marriage and divorce statistics

(i) Find for both distributions the median, lower quartile and upper quartile age at marriage.
(ii) Using the results of part (i) find appropriate measures and hence compare the two distributions. Comment on your results.

(ICSA June 1984)

(*Answers:* (i) males 23.98, 21.57, 27.78; females 22.11, 19.56, 24.52; (ii) males QD 3.1, skewness 0.22; females QD 2.5, skewness −0.03)

Q5

The data and parts (a) and (b) of this question were set as Exercises 1 and 2 of Chapter 4 (see Table 4.2).

Note: As there are a number of ways of doing part (a) there are a number of different answers to parts (c)–(e). Check the solution to Exercise 1 before proceeding (see Fig. 4S.1).

(c) Establish the value of the mode of the frequency distribution from the histogram.

(d) Establish the value of the mean of the distribution.

(e) Establish the value of the standard deviation of the distribution.

(f) Describe briefly the shape of the frequency distribution, using the values you obtain in (c), (d) and (e).

CACA(ACCA) Dec. 1983)

(*Answers:* 533; 532; 12.1)

Q6

(a) Using the figures given below calculate: (i) the range; (ii) the arithmetic mean; (iii) the median; (iv) the lower quartile; (v) the upper quartile; (vi) the quartile deviation (vii) the mean deviation; (viii) the standard deviation.

2	5	7	8	11	15	17	18	22	26	30	32	36	39
40	43	45	47	51	53	55	58	60	64	66			

(b) Explain the term 'measure of dispersion' and state briefly the advantages and disadvantages of using the following measures of dispersion: (i) range; (ii) quartile deviation; (iii) mean deviation; (iv) standard deviation.

(ICMA May 1984)

(*Answers:* 64; 34; 36; 16; 52; 18; 17.2; 19.63)

E. OUTLINE ANSWERS TO EXAM QUESTIONS

A1

Less than	Cumulative frequency
0	4
2	15
4	32
6	44
8	48
10	50

$$\frac{n}{2} = \frac{50}{2} = 25 \qquad \frac{n}{4} = \frac{50}{4} = 12.5$$

$$\frac{3 \times n}{4} = \frac{3 \times 50}{4} = 37.5$$

From graph (Fig. 5S.2), median = 3.2, $Q_1 = 1.6$, $Q_3 = 4.8$

$$\frac{Q_3 - Q_1}{2} = \frac{4.8 - 1.6}{2} = 1.6$$

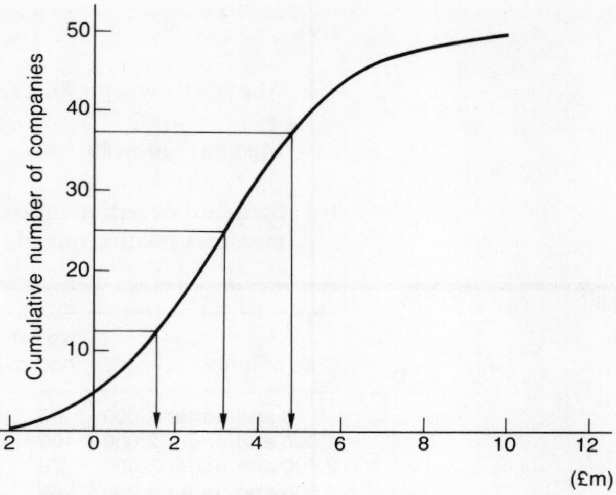

Fig. 5S.2 Cumulative number of companies

(iii) The median is a measure of central location. The median is the value such that half the values are greater than the median and half less. The quartile deviation is a measure of dispersion. The quartile deviation is equal to the average distance of the quartiles from the median.

A2

x	f	fx	fx^2	Less than	Cum. freq.
45	2	90	4,050	50	2
55	15	825	45,375	60	17
65	21	1,365	88,725	70	38
				←	M
75	30	2,250	168,750	80	68
85	20	1,700	144,500	90	88
95	9	855	81,225	100	97
105	3	315	33,075	110	100
Totals	100	7,400	565,700		

(i) $\bar{x}=\dfrac{\Sigma fx}{\Sigma f}=\dfrac{7,400}{100}=£74$

(ii) $\sigma=\sqrt{\dfrac{\Sigma fx^2}{\Sigma f}-\left(\dfrac{\Sigma fx}{\Sigma f}\right)^2}=\sqrt{\dfrac{565,700}{100}-\left(\dfrac{7,400}{100}\right)^2}=\sqrt{181}=£13.45$

(iii) Coefficient of variation

$$=\frac{\sigma\times 100}{\bar{x}}=\frac{13.45\times 100}{74}=18.2$$

(iv) $\dfrac{n}{2} = \dfrac{100}{2} = 50$.

Above arrow $L_1 = 70$, $\Sigma f_1 = 38$. Below arrow $L_2 = 80$, $\Sigma f_2 = 68$.

$$\text{Median} = 70 + (80 - 70)\dfrac{(50 - 38)}{(68 - 38)} = 70 + \dfrac{10 \times 12}{30} = £74$$

(b) Standard deviation measures spread, coefficient of variation measures relative spread.

A3

Size of loan	No. of borrowers	x (£'00)	f	fx	fx^2
0 and under 1,000	39	5	39	195	975
1,000 and under 2,000	100	15	100	1,500	22,500
2,000 and under 3,000	78	25	78	1,950	48,750
3,000 and under 4,000	42	35	42	1,470	51,450
4,000 and under 5,000	26	45	26	1,170	52,650
5,000 and under 6,000	15	55	15	825	45,375
Totals			300	7,110	221,700

$$\bar{x} = \dfrac{\Sigma fx}{\Sigma f} = \dfrac{7,110}{300} = 23.7$$

$$\sigma = \sqrt{\dfrac{\Sigma fx^2}{\Sigma f} - \left(\dfrac{\Sigma fx}{\Sigma f}\right)^2} = \sqrt{\dfrac{221,700}{300} - \left(\dfrac{7,110}{300}\right)^2} = \sqrt{177.31} = 13.3$$

To obtain results in correct units multiply by 100;
$\bar{x} = 23.7 \times 100 = £2,370 \qquad \sigma = 13.3 \times 100 = £1,330$

A4

(i) The question stated 'find', so graphical or calculation methods may be used. In this solution both methods have been used: calculation for males, graphical for females.

Less than	Cumulative frequency males	females		
20	24	77		
	← Q_1			
	← M			
25	166	219		
	← Q_3			
30	237	255		
35	260	266		
40	266	269		
45	271	272		
50	274	274		

$$\dfrac{n}{2} = \dfrac{274}{2} = 137 \qquad \dfrac{n}{4} = \dfrac{274}{4} = 68.5$$

$$\dfrac{3 \times n}{4} = \dfrac{3 \times 274}{4} = 205.5$$

Males
Above M arrow, $L_1 = 20$, $\Sigma f_1 = 24$. Below arrow, $L_2 = 25$, $\Sigma f_2 = 166$

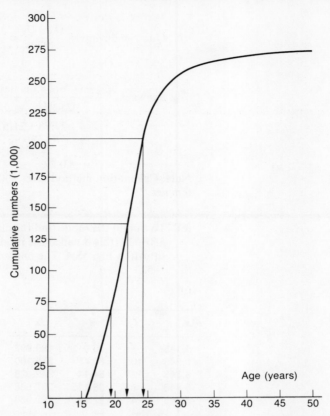

Fig. 5S.3   *Cumulative number of females, age when married*

Median $= 20 + (25 - 20)\dfrac{(137 - 24)}{(166 - 24)} = 20 + \dfrac{5 \times 113}{142} = 23.98$

Above Q_1 arrow, $L_1 = 20$, $\Sigma f_1 = 24$. Below arrow, $L_2 = 25$, $\Sigma f_2 = 166$

$Q_3 \quad = 20 + (25 - 20)\dfrac{(68.5 - 24)}{(166 - 24)} = 20 + \dfrac{5 \times 44.5}{142} = 21.57$

Above Q_3 arrow $L_1 = 25$, $\Sigma f_1 = 166$. Below arrow, $L_2 = 30$, $\Sigma f_2 = 237$

$Q_3 \quad = 25 + (30 - 25)\dfrac{(205.5 - 166)}{(237 - 166)} = 25 + \dfrac{5 \times 39.5}{71} = 27.78$

Females
From graph (Fig. 5S.3):
Median $= 22$
$Q_1 = 19.5$, $Q_3 = 24.5$

75

(ii) The question asked for 'appropriate measures'. As the plural is used you need more than one measure.

	Males	Females
Quartile deviation $\frac{Q_3 - Q_1}{2} =$	$\frac{27.78 - 21.57}{2}$	$\frac{24.5 - 19.5}{2}$
	$= 3.1$	$= 2.5$

Skewness $\dfrac{Q_3 + Q_1 - 2 \times \text{median}}{Q_3 - Q_1}$

$$= \frac{27.78 + 21.57 - 2 \times 23.98}{27.78 - 21.57} \quad\bigg|\quad \frac{24.5 + 19.5 - 2 \times 22}{24.5 - 19.5}$$

$$= 0.22 \qquad\qquad\qquad\qquad = 0$$

Note: Calculation method gives -0.03 as skewness and 2.48 as QD for females.

A5

(c) To obtain the mode find the class with largest frequency, namely 530–539 (this is called the modal group), and do the construction shown in Fig. 5S.4. The mode is read from the horizontal axes $= 533$.

(d, e)

x	f	fx	fx²
5.145	7	36.015	185.297
5.245	10	52.45	275.100
5.345	12	64.14	342.828
5.445	7	38.115	207.536
5.545	4	22.18	122.988
Totals	40	212.9	1,133.749

$$\bar{x} = \frac{\Sigma fx}{\Sigma f} = \frac{212.9}{40} = 5.32$$

$$\sigma = \sqrt{\frac{\Sigma fx^2}{\Sigma f} - \left(\frac{\Sigma fx}{\Sigma f}\right)^2}$$

$$= \sqrt{\frac{1,133.749}{40} - \left(\frac{212.9}{40}\right)^2}$$

$$= \sqrt{0.01472} = 0.121$$

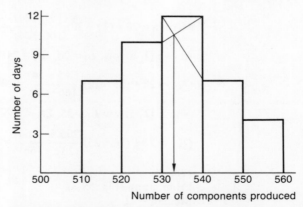

Fig. 5S.4   Distribution of number of components

To obtain results in correct units multiply by 100:
$$\bar{x} = 5.32 \times 100 = 532 \qquad \sigma = 0.121 \times 100 = 12.1$$

(f) To investigate shape we consider skewness. The mode is 533 and the mean is 532. These values are very close – the distribution is nearly symmetrical.

A6

(a)(i) Range $= 66 - 2 = 64$.

(ii) Total of all items $= 850$, mean $\dfrac{850}{25} = 34$.

(iii) As number of items is 25, median is 13th item $= 36$.

(iv) Q_1 is 6.5th item, i.e. the mean of 15 and $17 = 16$.

(v) Q_3 is 19.5th item, i.e. mean of 51 and $53 = 52$.

(vi) $\text{QD} = \dfrac{Q_3 - Q_1}{2} = \dfrac{52 - 16}{2} = 18$.

(vii) To find mean deviation subtract mean $= 34$ from each item and ignore sign: 32, 29, 27, 26, 23, 19, 17, 16, 12, 8, 4, 2, 2, 5, 6, 9, 11, 13, 17, 19, 21, 24, 26, 30, 32.

Total $= 430$; mean deviation is $\dfrac{430}{25} = 17.2$.

(viii) To find standard deviation square all the differences and add, i.e. $32 + 29 + \ldots + 30 + 32 = 9{,}636$.

Thus $\sigma = \sqrt{\dfrac{9{,}636}{25}} = \sqrt{385.44} = 19.63$.

(b) Measures of dispersion calculate 'spread' of data. Range is easy to find, useful in quality control. Not suitable if sample size is more than 10. Quartile deviation is relatively easy to understand and to find, useful for skew distributions. Ignores observations outside the middle 50%, not suitable for theoretical work. Mean deviation is easy to understand and to find. Not suitable for theoretical work. Standard deviation is the most important measure (see Chapter 11 for use in confidence intervals, etc.). Most difficult measure to calculate and understand.

A STEP FURTHER

Mulholland and Jones, *Fundamentals of Statistics*, Chs 5, 6 and 7.

Chapter 6

Regression and correlation

A. GETTING STARTED

We are often interested in finding whether a relationship exists between two variables. For example, we would expect there to be a relationship between advertising costs and sales; the more a company spends on advertising, the greater the sales it could expect. To investigate this relationship we could take a random sample of companies, and ask the companies to state the amount spent on advertising and the corresponding sales of the product. Suppose we obtained the data from 10 companies as in Table 6.1.

Worked Example 1

Table 6.1

Company	Sales (1,000)	Advertising costs (£100)
A	25	8
B	35	12
C	29	11
D	24	5
E	38	14
F	12	3
G	18	6
H	27	8
I	17	4
J	30	9

The first step is to plot the data in Table 6.1 on graph paper. The standard procedure is to plot the *dependent* variable on the *Y* axis (vertical) and the other variable on the *X* axis (horizontal). In the data above the dependent variable is sales – we are assuming that sales depend on advertising. The data is plotted in Fig. 6.1. The resulting graph is called a *scatter diagram*.

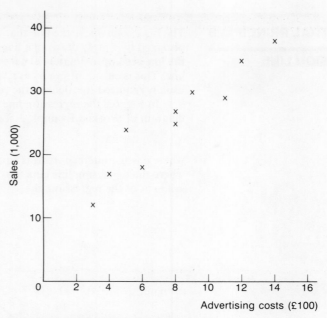

Fig. 6.1   Sales and advertising costs

The scatter diagram shows that there is a good relationship between sales and advertising costs. As sales increase with increasing advertising we say there is a positive relationship between the two variables. In Exercise 1 you will find a negative relationship.

Exercise 1

Table 6.2 shows numbers of cinema admissions and colour television licences.

Table 6.2

Year	Cinema admissions (millions)	Colour television licences (millions)
1973	134	5.0
1974	138	6.8
1975	116	8.3
1976	104	9.6
1977	103	10.7
1978	126	12.0
1979	112	12.7
1980	96	12.9

Draw a scatter diagram for the data in Table 6.2 by plotting cinema admissions and colour television licences.

The *regression line* is the line that best fits the data. This can be obtained by simply drawing a line so that the number of points 'above' the line are approximately equal to the number of points 'below' the line. This is called 'fitting by eye'. In examination questions you are usually required to calculate the equation of the line using a formula.

In Fig. 6.2 the regression line has been drawn for the scatter diagram of Worked Example 1. The equation of the regression line is:

$$Y = a + bX$$

where *a* is the intercept (this is equal to the distance between the point where the regression line cuts the *Y* axis and the origin), and *b* is the gradient of the regression line.

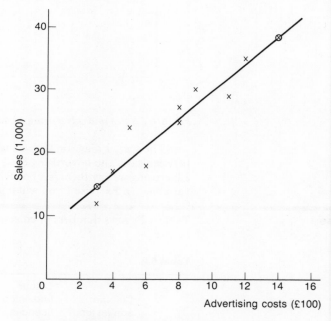

Fig. 6.2   Sales and advertising costs

The regression line equation $Y = a + bX$ is often called the *least squares regression line of Y on X*. This is because the regression line is that line for which the sum of the squared deviations between the observations and the line is 'the least' (i.e. a minimum). It is possible to show mathematically that the values *a* and *b* are given by the following formulae:

$$b = \frac{n\Sigma(YX) - (\Sigma Y) \times (\Sigma X)}{n\Sigma(X^2) - (\Sigma X)^2} \qquad a = \frac{\Sigma Y}{n} - b \times \frac{(\Sigma X)}{n}$$

where *n* is the number of pairs of observations.

There are other formulae used to find *a* and *b*, but the formulae above are the simplest to apply.

The value *b* is called the *regression coefficient*; it shows the change in *Y* if *X* changes by one unit. If the regression coefficient is 3, then if the value of *X* increases by 1, the value of *Y* increases by 3. The value of *b* can be positive or negative. If the regression line slopes upwards, *b* is positive; if the regression line slopes downwards, *b* is negative.

There are two regression lines, *Y* on *X* and *X* on *Y*. In examination questions you are usually asked to find the regression line of *Y* on *X*. However, you are not always told which set of data is the *Y* data and which set is the *X* data. It is essential to label the data correctly – you will lose marks if you find the wrong regression line. As explained on page 78, the *dependent* variable is the *Y* data. In some cases it is obvious which is the dependent variable and you should have no difficulty in labelling the two sets of data. Sometimes examination questions ask you to forecast a future value, e.g. to forecast future sales; in this case the sales data is the *Y* data. Sometimes you are asked to find the regression line of one variable on another, i.e. to find the regression line of sales on advertising costs; remember the regression is *Y* on *X*, thus sales are the *Y* data, advertising costs are the *X* data.

Worked Example 2

Using the data of Worked Example 1 (Table 6.1), find the equation of the regression line of sales on advertising costs.

Use the regression line to forecast sales if advertising costs were £1,000.

Here we are regressing sales on advertising costs, so sales are the *Y* data and advertising costs are the *X* data. We have 10 pairs of data so $n = 10$. We need to find ΣY, ΣX, ΣX^2, and $\Sigma(YX)$, as in Table 6.3.

Table 6.3

	Y	X	X²	YX
	25	8	64	200
	35	12	144	420
	29	11	121	319
	24	5	25	120
	38	14	196	532
	12	3	9	36
	18	6	36	108
	27	8	64	216
	17	4	16	68
	30	9	81	270
Totals	255	80	756	2,289

$$b = \frac{n\Sigma(YX) - (\Sigma Y) \times (\Sigma X)}{n\Sigma X^2 - (\Sigma X)^2} = \frac{10 \times 2,289 - 255 \times 80}{10 \times 756 - 80^2}$$

81

$$=\frac{22{,}890-20{,}400}{7{,}560-6{,}400}=\frac{2{,}490}{1{,}160}=2.14655.$$

$$a=\frac{\Sigma Y}{n}-b\times\frac{\Sigma X}{n}=\frac{255}{10}-2.14655\times\frac{80}{10}$$

$$=25.5-17.1724=8.3276$$

$$Y=a+bX \qquad Y=8.33+2.15X$$

Note: In examinations many candidates lose marks because of premature approximation. Above we found $b=2.14655$. If you were to approximate this to (say) 2.1 and then use this value to find a you would obtain $a=8.7$. **Always** use all places of decimals in intermediate calculations; at the **final** stage give result to appropriate accuracy.

To forecast sales if advertising costs were £1,000 we put $X=10$ in the equation (remember X is measured in 100's). We find

$$Y=8.33+2.15\times10=8.33+21.5=29.83$$

As the original data was given to the nearest integer (whole number) the forecast of sales $=30$ (or £30,000).

Examination questions sometimes expect you to plot the regression line on the scatter diagram. Take two values of X, one the smallest value of X the other the largest, and substitute in the regression equation:

$X=3 \qquad Y=8.33+2.15\times3\ =14.78$
$X=14 \qquad Y=8.33+2.15\times14=38.43$

Plot these two points on the scatter diagram, and join by a straight line. See Fig. 6.2.

Exercise 2

Table 6.4 shows the turnover and profit before taxation of Marks and Spencer plc from 1977 to 1982.

Table 6.4

Year	Turnover (£10m)	Profit before taxation (£10m)
1977	106	10
1978	125	12
1979	147	16
1980	167	17
1981	187	18
1982	220	22

Source: Company accounts.

(i) Plot a scatter diagram showing the relationship between profit before taxation and turnover.

(ii) Calculate the least squares regression line of profit before taxation on turnover.

(iii) Comment generally on your results.

(ICSA June 1984)

(*Answer:* $Y = -0.325 + 0.102X$)

USE OF AN ASSUMED MEAN

In some situations large numbers can cause calculator 'overflow'. Worked Example 3 shows how to deal with this problem.

Worked Example 3

Table 6.5 shows the increase in average earnings of male employees in the United Kingdom between 1975 and 1981.

Table 6.5

Year	Average earnings (£)
1975	59
1976	70
1977	77
1978	87
1979	99
1980	122
1981	137

(i) Find the equation of the least squares regression line which would enable you to forecast earnings in future years.

(ii) Make a forecast of the earnings for 1982.

Average earnings are the Y data, years the X data. With an odd number of years the simplest procedure is to make the middle year (1978) equal to zero; if there are an even number of years the easiest way is to make the first year equal to zero or unity.

(i)

X	Y	X^2	YX
-3	59	9	-177
-2	70	4	-140
-1	77	1	-77
0	87	0	0
1	99	1	99
2	122	4	244
3	137	9	411
Totals 0	651	28	360

$$b = \frac{n\Sigma(YX) - (\Sigma Y) \times (\Sigma X)}{n\Sigma X^2 - (\Sigma X)^2} = \frac{7 \times 360 - 651 \times 0}{7 \times 28 - 0}$$

$$= \frac{2,520}{196} = 12.8571$$

$$a = \frac{\Sigma Y}{n} - b \times \frac{\Sigma X}{n} = \frac{651}{7} - 12.8571 \times \frac{0}{7} = 93$$

$$Y = a + bX \qquad Y = 93 + 12.86X$$

(ii) To make a forecast for 1982 put $X = 4$ in the equation (if 1978 = 0 then 1982 = 4).

$$Y = 93 + 12.8571 \times 4 = 93 + 51.4284 = 144.4284 = 144 = \pounds 144.$$

Exercise 3

A trading corporation obtained the profits after tax shown in Table 6.6 for the years 1975 to 1981 inclusive.

Table 6.6

Year	Profits after tax (£'000)
1976	261
1976	273
1977	299
1978	312
1979	327
1980	341
1981	350

(i) Find the trend using the method of least squares, and enter the data on a historigram, superimposing the trend thus calculated.

(ii) Estimate profits after tax for 1982, and briefly discuss the likely accuracy of your estimate.

(ICSA June 1982)

(*Answers:* $Y = 309 + 15.39X$ (1978 = 0); 371 = £371,000)

CORRELATION

The regression equation is the best fitting line to the data. In Fig. 6.3 the regression line would clearly be a good fit to the data, whereas in Fig. 6.4 the regression line would be a poor fit to the data.

To find whether a regression line is a good fit to the data, we could draw a scatter diagram. In an examination situation, however, you should draw a scatter diagram only if asked to do so. It is possible to find whether a regression line is a good fit to data by calculating a *correlation coefficient*. If the correlation coefficient is $+1$, then the regression line fits the data perfectly i.e. all the data lies on a regression line. As the coefficient is positive this means the regression line slopes upwards. If the correlation coefficient is -1, the regression line fits the data perfectly but the regression line slopes downwards. In Fig. 6.3 the

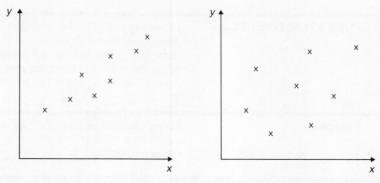

Fig. 6.3   Fig. 6.4

correlation coefficient would be about $+0.9$, in Fig. 6.4 about $+0.3$. A correlation coefficient of 0 means that there is no relationship between the two sets of data.

PRODUCT MOMENT CORRELATION COEFFICIENT

The formula for this coefficient is shown below:

$$r = \frac{n\Sigma(YX) - (\Sigma Y) \times (\Sigma X)}{\sqrt{(n\Sigma X^2 - (\Sigma X)^2)(n\Sigma Y^2 - (\Sigma Y)^2)}}$$

The value of r must lie between -1 and $+1$. If you obtain a value of r numerically greater than 1, you have made a mistake, so carefully check you have the correct formula and check your arithmetic.

When we found the regression line we had to calculate ΣX, ΣY, ΣX^2, and $\Sigma(YX)$. To find a product moment correlation coefficient we have to find ΣY^2 in addition.

Worked Example 4

Using the data of Worked Example 1 calculate the product moment correlation coefficient (see Table 6.7).

Table 6.7

Y	X	Y²	X²	YX
25	8	625	64	200
35	12	1,225	144	420
29	11	841	121	319
24	5	576	25	120
38	14	1,444	196	532
12	3	144	9	36
18	6	324	36	108
27	8	729	64	216
17	4	289	16	68
30	9	900	81	270
255	80	7,097	756	2,289

$$r = \frac{n\Sigma(YX) - (\Sigma Y) \times (\Sigma X)}{\sqrt{(n\Sigma X^2 - (\Sigma X)^2)(n\Sigma Y^2 - (\Sigma Y)^2)}}$$

$$r = \frac{10 \times 2,289 - 255 \times 80}{\sqrt{(10 \times 756 - 80^2)(10 \times 7,097 - 255^2)}}$$

$$r = \frac{22,890 - 20,400}{\sqrt{(7,560 - 6,400)(70,970 - 65,025)}}$$

$$r = \frac{2,490}{\sqrt{1,160 \times 5,945}} = \frac{2,490}{\sqrt{6,896,200}} = \frac{2,490}{2,626.06} = 0.948$$

In working out the denominator of a correlation coefficient, it is possible to have calculator overflow on inexpensive calculators. To show how to overcome this problem, suppose we need the square root of $21,360 \times 62,581$. Find the square root of $21,360 = 146.15061$ and square root of $62,581 = 250.16195$, and then multiply 146.15061 and $250.16195 = 36561.321$.

Exercise 4

Using the data of Exercise 2 calculate the product moment correlation coefficient.

(*Answer:* 0.985)

RANK CORRELATION COEFFICIENT

In some situations data is given ranked, for example, applicants for a post may be interviewed by two senior staff. Each interviewer puts the applicants in an order of merit. We may wish to examine the degree of agreement of the two interviewers. We could calculate the product moment correlation coefficient, but there is a quicker procedure. We use the *rank correlation coefficient*. (This coefficient is often called *Spearman's* rank correlation coefficient.) The formula for the rank correlation coefficient is as follows:

$$r = 1 - \frac{6\Sigma d^2}{n^3 - n}$$

where d is the difference in the ranks and n is the number of observations.

Worked Example 5

Ten candidates for an administrative post were ranked by the two members of the interviewing panel in the following manner:

		Candidates (duly ranked)									
		A	B	C	D	E	F	G	H	I	J
Panel	I	4	2	7	1	5	6	9	3	10	8
members	II	3	2	5	1	4	9	6	7	8	10

Calculate Spearman's rank correlation coefficient and discuss whether it represents a measure of agreement between the two panel members.

(LCCI May 1983)

The first step is to take the difference in the ranks; the second step is to square each of these differences; the third step is to sum the squares of the differences; finally substitute in the formula.

d	1	0	2	0	1	3	3	4	2	2
d^2	1	0	4	0	1	9	9	16	4	4

$\Sigma d^2 = 48$

$$r = 1 - \frac{6\Sigma d^2}{n^3 - n} = 1 - \frac{6 \times 48}{10^3 - 10} = 1 - \frac{288}{990} = 0.709$$

The correlation coefficient of 0.709 shows that there is some degree of agreement of the panel members.

Exercise 5

The personnel department of a large company is investigating the possibility of assessing the suitability of applicants by using psychological tests instead of normal interview procedures. A comparative test of seven applicants was carried out using both methods. The results were as shown in Table 6.8.

Table 6.8

Applicant	Ranking by interview procedure	Ranking by psychological tests
A	4	5
B	1	2
C	7	7
D	6	4
E	2	1
F	3	3
G	5	6

(i)   Calculate the rank coefficient of correlation.
(ii) Interpret the result established.
(*Answer:* 0.857)

It is possible to calculate a rank correlation coefficient when data is given unranked. In this situation it is first necessary to rank the data. The procedure to follow is illustrated in Worked Example 6.

Worked Example 6

Table 6.9 shows sales of gas and electricity by the public supply system.

Table 6.9

	Gas (1,000 million therms)	Electricity (10,000 GWh)
1974	12.7	21.9
1975	13.1	21.8
1976	14.0	22.1
1977	14.6	22.7
1978	15.3	23.0
1979	16.5	24.1
1980	16.7	22.9
1981	16.6	22.7

Rank these data and calculate a coefficient of correlation.

The first step is to rank the data for each series, give rank 1 to the largest observation, rank 2 to the next largest and so on. When there is a tie, as there is for the electricity data (1977 and 1981 sales are the same), the simplest procedure is to take the ranks that 22.7 would have, 4 and 5, take the average 4.5, and give this as the ranks for 1977 and 1981. The result is shown in Table 6.10.

Table 6.10

	1974	1975	1976	1977	1978	1979	1980	1981
Gas	8	7	6	5	4	3	1	2
Electricity	7	8	6	4.5	2	1	3	4.5
d	1	1	0	0.5	2	2	2	2.5
d^2	1	1	0	0.25	4	4	4	6.25

$$\Sigma d^2 = 20.5$$

$$r = 1 - \frac{6\Sigma d^2}{n^3 - n} = 1 - \frac{6 \times 20.5}{8^3 - 8} = 1 - \frac{123}{504} = 0.756$$

Exercise 6

The data in Table 6.11 gives the actual sales of a company in each of eight regions of a country together with the forecast of sales by two different methods.

Table 6.11

Region	Actual sales	Forecast 1	Forecast 2
A	15	13	16
B	19	25	19
C	30	23	26
D	12	26	14
E	58	48	65
F	10	15	19
G	23	28	27
H	17	10	22

(i)   Calculate the rank correlation coefficient between
      (1) Actual sales and forecast 1
      (2) Actual sales and forecast 2.
(ii) Which forecasting method would you recommend next year?

(ICSA June 1983)

(*Answers:* 0.524; 0.839)

COEFFICIENT OF DETERMINATION

The dependent variable is Y; X is often called the *explanatory variable*. In Worked Example 1 sales were the dependent variable and advertising costs the explanatory variable. The *coefficient of determination*, which is equal to r^2, the square of the correlation coefficient, shows how much of the variation of the dependent variable can be 'explained' by the explanatory variable. In Worked Example 4 we found that the correlation coefficient (r) for the sales–advertising costs data was 0.948. If we square r we obtain 0.899; thus 89.9% of the variation of sales can be 'explained' by advertising costs.

Low values of r should be interpreted with caution. If $r = 0.4$ then $r^2 = 0.16$, i.e. only 16% of the variation of Y can be 'explained' by X. Note that for $r = 0.7$, $r^2 = 0.49$; thus for even a fairly high value of r, less than half the variation of Y is 'explained' by X.

C. SOLUTIONS TO EXERCISES

S1

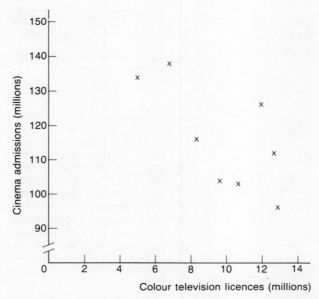

Fig. 6S.1   Cinema admissions and colour television licences

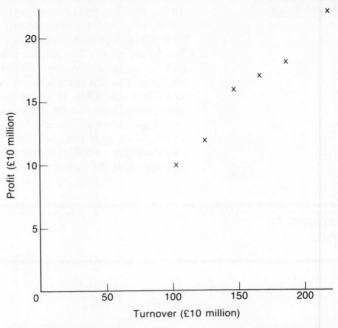

Fig. 6S.2   Marks & Spencer plc profit before taxation and turnover
Source: company accounts

Y	X	X²	Y²	XY
10	106	11,236	100	1,060
12	125	15,625	144	1,500
16	147	21,609	256	2,352
17	167	27,889	289	2,839
18	187	34,969	324	3,366
22	220	48,400	484	4,840
95	952	159,728	1,597	15,957

$$b = \frac{n\Sigma(YX) - (\Sigma Y) \times (\Sigma X)}{n\Sigma X^2 - (\Sigma X)^2}$$

$$= \frac{6 \times 15,957 - 95 \times 952}{6 \times 159,728 - 952^2}$$

$$= \frac{95,742 - 90,440}{958,368 - 906,304} = \frac{5,302}{52,064}$$

$$= 0.10184$$

$$a = \frac{\Sigma Y}{n} - b \times \frac{\Sigma X}{n}$$

$$= \frac{95}{6} - 0.10184 \times \frac{952}{6}$$

$$= 15.8333 - 16.158 = -0.325$$

$$Y = -0.325 + 0.102X$$

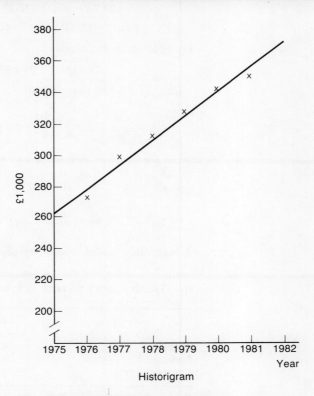

Historigram

Fig. 6S.3   Profits after tax 1975–81

Y	X	X²	YX
261	−3	9	−783
273	−2	4	−546
299	−1	1	−299
312	0	0	0
327	1	1	327
341	2	4	682
350	3	9	1,050
2,163	0	28	431

$$b = \frac{n\Sigma(YX) - (\Sigma Y) \times (\Sigma X)}{n\Sigma X^2 - (\Sigma X)^2}$$

$$= \frac{7 \times 431 - 2,163 \times 0}{7 \times 28 - 0}$$

$$= \frac{3,017}{196} = 15.3929$$

$$a = \frac{\Sigma Y}{n} - b \times \frac{\Sigma X}{n} = \frac{2,163}{7} - 15.3929 \times \frac{0}{7} = 309$$

$$Y = 309 + 15.39X$$

For 1982, $X = 4$ $\therefore$ $T = 309 + 15.3929 \times 4 = 370.57 = 371 = £371,000$

S4

$$r = \frac{n\Sigma(YX) - (\Sigma Y) \times (\Sigma X)}{\sqrt{(n\Sigma X^2 - (\Sigma X)^2)(n\Sigma Y^2 - (\Sigma Y)^2)}}$$

Using totals from Exercise 2,

$$r = \frac{6 \times 15,957 - 95 \times 952}{\sqrt{(6 \times 159,728 - 952^2)(6 \times 1,597 - 95^2)}} = \frac{5,302}{\sqrt{52,064 \times 557}}$$

$$= \frac{5,302}{\sqrt{28,999,648}} = \frac{5,302}{5,385.13} = 0.985$$

S5

d	1	1	0	2	1	0	1
d^2	1	1	0	4	1	0	1

$$\Sigma d^2 = 8 \qquad r = 1 - \frac{6 \times \Sigma d^2}{n^3 - n} = 1 - \frac{6 \times 8}{7^3 - 7} = 1 - 0.1429 = 0.857$$

A high value, and thus good agreement between interviewers.

S6

(i)   The first step is to rank each set of data.

	A	B	C	D	E	F	G	H
Sales	6	4	2	7	1	8	3	5
Forecast 1	7	4	5	3	1	6	2	8
Forecast 2	7	5.5	3	8	1	5.5	2	4

Sales − Forecast 1		A	B	C	D	E	F	G	H
	d	1	0	3	4	0	2	1	3
	d^2	1	0	9	16	0	4	1	9

$$\Sigma d^2 = 40 \qquad r = 1 - \frac{6 \times \Sigma d^2}{n^3 - n} = 1 - \frac{6 \times 40}{8^3 - 8} = 1 - 0.476 = 0.524$$

Sales − Forecast 2		A	B	C	D	E	F	G	H
	d	1	1.5	1	1	0	2.5	1	1
	d^2	1	2.25	1	1	0	6.25	1	1

$$\Sigma d^2 = 13.5 \qquad r = 1 - \frac{6 \times \Sigma d^2}{n^3 - n} = 1 - \frac{6 \times 13.5}{8^3 - 8} = 1 - 0.161 = 0.839$$

(ii)   There was a closer agreement between sales and forecast 2; therefore recommend forecast method 2 next year.

D. RECENT EXAMINATION
QUESTIONS

Q1

Ten printers suitable for use with microcomputers and retailing at between £350 and £550 were evaluated by a group of twenty-five computer owners. They were asked to award each printer marks out of ten for 'speed' and 'print quality'. Table Q1 summarizes their assessments.

Table Q1

Model	Speed (average marks out of 10)	Print quality (average marks out of 10)	Overall ranking	Retail price (£)
1	4	7	5	430
2	9	4	3	416
3	5	6	8	350
4	2	9	7	550
5	6	2	4	419
6	5	3	10	373
7	7	5	1	450
8	6	3	2	424
9	4	6	8	370
10	5	4	6	399

(a) Calculate the product moment coefficient of correlation between 'speed' and 'print quality' and interpret your answer.

(b) Calculate the rank correlation coefficient between 'overall ranking' and 'retail price' and interpret your answer.

(ABE Dec. 1983)

(*Answers:* -0.654; 0.621)

Q2

A medium-sized manufacturing organization has the following sales analysis for the period 1971–80 inclusive.

Year	1971	1972	1973	1974	1975	1976	1977	1978	1979	1980
Sales (£m.)	15.3	14.6	16.8	17.3	17.2	20.9	22.3	20.0	23.1	24.5

(i) Calculate the trend line using the least squares equation.

(ii) Calculate the trend values for 1971 and 1980 and explain the meaning of these results.

(iii) Draw a graph of the above data including the trend and project the trend for the year 1981.

(ICMA Nov. 1980)

(*Answers:* $Y = 13.3 + 1.073X$ (1971 = 1); 14.4; 24.0; 25.1 = £25.1m.)

Q3

(a) 'Correlation does not prove causation.' Discuss this statement.

(b) Table Q3 shows cinema admissions and colour television licences.

Table Q3

Year	1973	1974	1975	1976	1977	1978	1979	1980
Cinema admissions (m.)	5.0	6.8	8.3	9.6	10.7	12.0	12.7	12.9
Colour TV licences (m.)	134	138	116	104	103	126	112	96

Calculate the product moment correlation coefficient between the number of cinema admissions and the number of colour television licences issued. Briefly comment on your results.

(ICSA Dec. 1983)

(*Answer:* -0.692)

Q4

Three sales executives have been asked to rank the main styles of product of a shoe manufacturing firm in order of merit for a new market. The results of the judging are shown in Table Q4.

Table Q4

Sales executive	Product style									
	I	*II*	*III*	*IV*	*V*	*VI*	*VII*	*VIII*	*IX*	*X*
A	5	10	1	6	2	7	8	9	3	4
B	5	8	4	9	1	6	10	7	3	2
C	7	8	1	10	2	5	9	6	4	3

Using rank correlation coefficients, discuss the nature of agreement between the executives. Discuss your results.

(ICSA Dec. 1979)

(*Answers:* AB 0.782; BC 0.879; AC 0.758)

Q5

The following table shows the number of staff in each of six regional offices and the total running costs (including salaries) of these offices:

Number of staff	10	8	14	16	11	7
Running costs (£1000)	102	93	143	134	121	99

(i) Draw a suitable graph to illustrate this information.

(ii) Calculate the equation of the line of regression of total running costs on number of staff and draw it on your graph.

(iii) Predict the running costs of a regional office employing 12 staff.

(LCCI May 1984)

(*Answers:* $Y = 56.3 + 5.367X$; $120.7 = £120,700$)

E. OUTLINE ANSWERS TO EXAM QUESTIONS

A1

(a)

X	Y	X²	Y²	XY
7	4	49	16	28
4	9	16	81	36
6	5	36	25	30
9	2	81	4	18
2	6	4	36	12
3	5	9	25	15
5	7	25	49	35
3	6	9	36	18
6	4	36	16	24
4	5	16	25	20
49	53	281	313	236

$$r = \frac{10 \times 236 - 53 \times 49}{\sqrt{(10 \times 281 - 49^2)(10 \times 313 - 53^2)}}$$

$$= \frac{-237}{\sqrt{409 \times 321}}$$

$$= \frac{-237}{362.34} = -0.654$$

Negative relationship – the faster the print speed, the lower the print quality.

(b) The data needs to be ranked:

Ranking	5	3	8.5	7	4	10	1	2	8.5	6
Price	3	6	10	1	5	8	2	4	9	7
d	2	3	1.5	6	1	2	1	2	0.5	1
d^2	4	9	2.25	36	1	4	1	4	0.25	1

$$\Sigma d^2 = 62.5 \qquad r = 1 - \frac{6 \times \Sigma d^2}{n^3 - n} = 1 - \frac{6 \times 62.5}{10^3 - 10} = 1 - 0.3788 = 0.621$$

Positive relationship – the higher the price, the higher the ranking.

A2

(i)

	X	Y	X²	YX
	1	15.3	1	15.3
	2	14.6	4	29.2
	3	16.8	9	50.4
	4	17.3	16	69.2
	5	17.2	25	86.0
	6	20.9	36	125.4
	7	22.3	49	156.1
	8	20.0	64	160.0
	9	23.1	81	207.9
	10	24.5	100	245.0
Totals	55	192.0	385	1,144.5

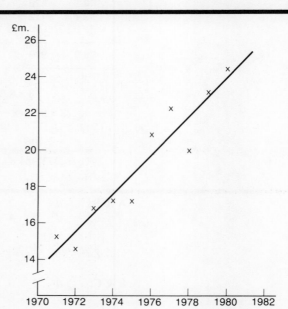

Fig. 6S.4   Sales 1971–80

$$b = \frac{10 \times 1,144.5 - 192 \times 55}{10 \times 385 - 55^2}$$

$$= \frac{885}{825} = 1.07273$$

$$a = \frac{192}{10} - 1.07273 \times \frac{55}{10} = 19.2 - 5.9 = 13.3$$

$$Y = 13.3 + 1.0727X$$

(ii) 1971: $X = 1$ $Y = 13.3 + 1.0727 \times 1 = 14.4$
    1980: $X = 10$ $Y = 13.3 + 1.0727 \times 10 = 24.0$
(iii) 1981: $X = 11$ $Y = 13.3 + 1.0727 \times 11 = 25.1 = £25.1\text{m}.$

A3

(a) The correlation coefficient shows whether a straight line is a good fit to the data. The fact that the correlation coefficient may be near to 1 merely shows that as the X values increase, so do the Y values; it does **not** follow that there is a causative relationship between the two variables.

(b)

X	Y	X²	Y²	YX
5.0	134	25.00	17,956	670.0
6.8	138	46.24	19,044	938.4
8.3	116	68.89	13,456	962.8
9.6	104	92.16	10,816	998.4
10.7	103	114.49	10,609	1,102.1
12.0	126	144.00	15,876	1,512.0
12.7	112	161.29	12,544	1,422.4
12.9	96	166.41	9,216	1,238.4
78.0	929	818.48	109,517	8,844.5

$$r = \frac{8 \times 8,844.5 - 929 \times 78}{\sqrt{(8 \times 818.48 - 78^2)(8 \times 109,517 - 929^2)}} = \frac{-1,706}{\sqrt{13,095 \times 463.84}}$$

$$= \frac{-1,706}{\sqrt{6,073,984.8}} = \frac{-1,706}{2,464.55} = -0.692$$

A4

A & B:

d	0	2	3	3	1	1	2	2	0	2
d²	0	4	9	9	1	1	4	4	0	4

$$\Sigma d^2 = 36 \quad r = 1 - \frac{6 \times \Sigma d^2}{n^3 - n} = 1 - \frac{6 \times 36}{10^3 - 10} = 1 - 0.2182 = 0.782$$

B & C:

d	2	0	3	1	1	1	1	1	1	1
d²	4	0	9	1	1	1	1	1	1	1

$$\Sigma d^2 = 20 \qquad r = 1 - \frac{6 \times \Sigma d^2}{n^3 - n} = 1 - \frac{6 \times 20}{10^3 - 10} = 1 - 0.1212 = 0.879$$

A & C:

d	2	2	0	4	0	2	1	3	1	1
d^2	4	4	0	16	0	4	1	9	1	1

$$\Sigma d^2 = 40 \qquad r = 1 - \frac{6 \times \Sigma d^2}{n^3 - n} = 1 - \frac{6 \times 40}{10^3 - 10} = 1 - 0.2424 = 0.758$$

A5

(i)

X	Y	X^2	YX
10	102	100	1,020
8	93	64	744
14	143	196	2,002
16	134	256	2,144
11	121	121	1,331
7	99	49	693
66	692	786	7,934

$$b = \frac{6 \times 7,934 - 692 \times 66}{6 \times 786 - 66^2} = \frac{1,932}{360}$$

$$= 5.36667$$

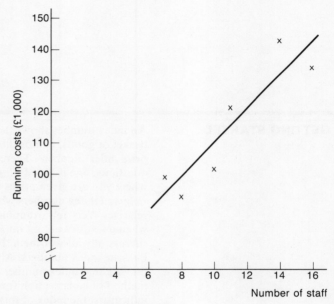

Fig. 6S.5   *Running costs*

$$a = \frac{692}{6} - 5.36667 \times \frac{66}{6}$$

$$= 115.333 - 59.033 = 56.30$$

$$Y = 56.30 + 5.367X$$

To plot the line take two values of X, one at each end.

$X = 7 \qquad Y = 56.30 + 5.36667 \times 7 = 93.9$

$X = 16 \qquad Y = 56.30 + 5.36667 \times 16 = 142.2$

(iii) $X = 12 \qquad Y = 56.30 + 5.36667 \times 12 = 120.7 = £120,700$

A STEP FURTHER   Mulholland and Jones, *Fundamentals of Statistics*, Ch. 13.

Chapter 7

Index numbers

A. GETTING STARTED

An index number shows the change in the value, quantity or price of a basket of goods. Almost all examination questions are involved with price index numbers. There are essentially two types of examination questions: one is based on *Laspeyres* and *Paasche* index numbers where you are given prices and quantities, the other is based on *weighted* index numbers where you are given weights and price relatives. Very few examination questions are set on the latter, and when questions are set on weighted index numbers, few candidates successfully answer them. It is essential to master Laspeyres and Paasche index numbers. Although less essential the part of the chapter on weighted index numbers should still be read, since this shows the method of construction for two major economic and business indicators: the index of retail prices and the index of industrial production.

B. ESSENTIAL PRINCIPLES

LASPEYRES AND PAASCHE INDICES

The *Laspeyres* price index takes a basket of goods purchased at some time in the past – the *base year* – and compares the cost of the basket of goods at *base year* prices with the cost of the same basket of goods at the prices prevailing in a *subsequent year*. A Laspeyres index is sometimes called a 'base weighted' index.

The disadvantage of the Laspeyres index is that as tastes change, the base year basket of goods may not be representative of the goods purchased today. To overcome this difficulty the *Paasche* index may be used. This index compares the cost of the basket of goods relevant to *today* at *base year* prices, with the cost of the same basket of goods at *today's* prices. A Paasche index is sometimes called a 'current weighted' index.

There are four formulae involved with Laspeyres and Paasche

index numbers. The first two are *price* index numbers and are the most important in terms of examination questions set; the second two are *quantity* index numbers, and are less often the basis of examination questions.

If p_0 is the base year price and q_0 is the base year quantity, and if p_n is the current year price and q_n is the current year quantity, then the formulae are as follows:

Laspeyres price index number	$\dfrac{\Sigma p_n q_0}{\Sigma p_0 q_0} \times 100$
Paasche price index number	$\dfrac{\Sigma p_n q_n}{\Sigma p_0 q_n} \times 100$
Laspeyres quantity index number	$\dfrac{\Sigma p_0 q_n}{\Sigma p_0 q_0} \times 100$
Paasche quantity index number	$\dfrac{\Sigma p_n q_n}{\Sigma p_n q_0} \times 100$

PRICE INDEX NUMBERS

Worked Example 1

(a) Discuss the relative advantages and disadvantages of Laspeyres and Paasche price index numbers.
(b) An investor's holding in shares of three companies in 1977 and 1982 is shown in Table 7.1. The average price of these shares in each year is also given.

Table 7.1

| Share | 1977 | | 1982 | |
	Price	Number of shares	Price	Number of shares
A	145	300	420	600
B	280	500	130	200
C	205	400	240	400

(i) Calculate Laspeyres and Paasche price index numbers for the investor's holding of shares with 1977 as base year.
(ii) Explain briefly why your results in part (i) differ.

(ICSA Dec. 1983)

(a) **Laspeyres price index number**
Advantages:
(i) The quantities have to be found for the base year only.
(ii) As the basket of goods is the same for each year, comparisons can be made between years.
(iii) Most index numbers in current use are of this type.
Disadvantages:

(i) As tastes change the base year quantities may not represent current consumption.

(ii) The Laspeyres index overstates price rises.

Paasche price index number

Advantages:

(i) Current year quantities represent the present consumption.

Disadvantages:

(i) Comparisons can be made only with the base year.

(ii) Quantities have to be found for each year.

(iii) In practice it is difficult to find current year quantities.

(iv) The Paasche index understates price rises.

(b) The first step is to label 1977 year prices and quantities p_0 and q_0 respectively, and 1982 prices and quantities p_n and q_n. For each item we need to find p_0q_0, p_nq_0, p_0q_n and p_nq_n, as in Table 7.2.

Table 7.2

Share	p_0	q_0	p_n	q_n	p_0q_0	p_nq_0	p_0q_n	p_nq_n
A	145	300	420	600	43,500	126,000	87,000	252,000
B	280	500	130	200	140,000	65,000	56,000	26,000
C	205	400	240	400	82,000	96,000	82,000	96,000
Totals					265,500	287,000	225,000	374,000

Laspeyres price index

$$= \frac{\Sigma p_n q_0}{\Sigma p_0 q_0} \times 100 = \frac{287,000}{265,500} \times 100 = 108.1$$

Paasche price index

$$= \frac{\Sigma p_n q_n}{\Sigma p_0 q_n} \times 100 = \frac{374,000}{225,000} \times 100 = 166.2$$

The large difference is due to the investor switching from shares which have fallen in value to shares which have risen in value.

Exercise 1

A company making components for the motor industry uses three main materials – plastic, steel tubing and cloth. Table 7.3 shows the price (in £) and the quantity used in each of the years 1979 and 1981.

Table 7.3

	Plastic		Steel tubing		Cloth	
	Price (£)	Quantity	Price (£)	Quantity	Price (£)	Quantity
1979	1.20	4,000	5.30	1,000	2.30	2,000
1981	2.50	2,000	5.80	800	2.70	4,000

(i) Calculate Laspeyres and Paasche price index numbers for 1981 with 1979 as base.

(ii) Interpret the results and explain why they are different.

(LCCI May 1984)

(*Answers:* 144.2; 129.0)

| QUANTITY INDEX NUMBERS | In Worked Example 1 *price* index numbers were found; in Worked Example 2 *quantity* index numbers will be found. |

Worked Example 2

Using the data of Worked Example 1, find Laspeyres and Paasche *quantity* index numbers.

All the necessary totals were found in Worked Example 1. Substituting in the formulae we find:

Laspeyres quantity index

$$= \frac{\Sigma p_0 q_n}{\Sigma p_0 q_0} \times 100 = \frac{225,000}{265,500} \times 100 = 84.7$$

Paasche quantity index

$$= \frac{\Sigma p_n q_n}{\Sigma p_n q_0} \times 100 = \frac{374,000}{287,000} \times 100 = 130.3$$

Exercise 2

Using the data of Exercise 1, find Laspeyres and Paasche *quantity* index numbers.

(*Answers:* 107.8; 96.4)

WEIGHTED INDEX NUMBERS

Laspeyres and Paasche index numbers assume that there is an identifiable unit of quantity. This is possible for food and drink but for housing there is no obvious unit of quantity. The standard practice is to give a 'weight' to each item which is proportional to the expenditure on that item at base year.

In the case of a *price* index, e.g. the index of retail prices, we find the *price relative* for each item. If p_0 is the price at base year and p_n is the price today, then the *price relative* is equal to:

$$\text{Price relative} = I = \frac{p_n}{p_0}$$

In the case of a *quantity* index, e.g. the index of industrial production, we find the *quantity relative* for each item. If q_0 is the quantity at base year and q_n is the quantity today, then the *quantity relative* is equal to:

$$\text{Quantity relative} = I = \frac{q_n}{q_0}$$

The formula for finding a *weighted index number* is as follows:

$$\text{Weighted index number} = \frac{\Sigma WI}{\Sigma W}$$

where W is the weight for the item and I is the price (or quantity) relative. It should be noted that the same formula is used for both price and quantity index numbers.

Worked Example 3

Table 7.4 shows the general index of retail prices at September 1983.

Table 7.4

Group	Weight	Price relatives (Jan. 1974 = 100)
Food	203	313.0
Alcoholic drink	78	371.8
Tobacco	39	443.5
Housing	137	376.7
Fuel and light	69	466.0
Durable household goods	64	251.6
Clothing and footware	74	215.8
Transport and vehicles	159	373.1
Miscellaneous goods	75	348.6
Services	63	344.7
Meals bought and consumed outside the home	39	368.9

Calculate the general index of retail prices for (i) all items; (ii) non-food items.

We need to find ΣW and ΣWI; this is best done in a tabulation, as in Table 7.5.

Table 7.5

W	I	WI
203	313.0	63,539.0
78	371.8	29,000.4
39	443.5	17,296.5
137	376.7	51,607.9
69	466.0	32,154.0
64	251.6	16,102.4
74	215.8	15,969.2
159	373.1	59,322.9
75	348.6	26,145.0
63	344.7	21,716.1
39	368.9	14,387.1
Totals 1,000		347,240.5

(i) General index of retail prices

$$= \frac{\Sigma WI}{\Sigma W} = \frac{347,240.5}{1,000} = 347.2$$

(ii) For non-food items we need to exclude the first and last entries. There are two methods of calculation: either add up the middle 9 entries, or subtract the first and last entries from the totals found in part (i). The following calculation uses the second method. Index for non-food items

$$= \frac{\Sigma WI}{\Sigma W} = \frac{347,240.5 - 63,539.0 - 14,387.1}{1,000 - 203 - 39}$$

$$= \frac{269,314.4}{758} = 355.3$$

The Managing Director of your company is concerned that between 1975 and 1983 the output of the company has increased by only 24%. He asks you to find out the increase in output for production industries throughout the UK. You obtain the information from the *Monthly Digest of Statistics* relating to August 1983 (Table 7.6).

Table 7.6　Index of industrial production

Group	Weight	Index (1980 = 100)
Energy and water supply	264	101
Manufacturing industries		
Metals	25	82
Mineral products	41	88
Chemicals	68	95
Engineering	325	84
Food, drink and tobacco	99	92
Textiles	52	82
Other	126	83

Calculate the index of industrial production for August 1983 with 1980 = 100 for: (i) all industries; (ii) all manufacturing industries.

(ICSA (part) Dec. 1984)

(*Answers:* 89.9; 85.9)

CHANGE OF BASE YEAR

From time to time the base year of index numbers has to change. In Exam Question 3 below, a question is set on the index of industrial production where the base is 1975 = 100. The weights for this index relate to the value of output in 1975. North Sea oil came on stream after 1975 and was, therefore, excluded from the weighting. The revised index of industrial production (which was the basis of Exercise 3 above) is based on 1980 = 100, and includes a weighting for North Sea oil. Sometimes it is desirable to make comparisons over a long period of time, and when there is a change of base within this period appropriate calculations have to be undertaken. How to do this is explained in Worked Example 4.

Worked Example 4

The general index of retail prices for January 1974 with January 1962 = 100 was 191.8. Calculate the general index of retail prices for September 1983 with January 1962 = 100.

In Worked Example 3 we found that the general index of retail prices for September 1983 with January 1974 = 100 was 347.2. We set out the calculations for a change of base year as follows.

January 1962	100	
January 1974	191.8	100
September 1983	X	347.2

We need to find the value X so that we can find the increase in prices relative to January 1962. From January 1974 prices increased from 100 to 347.2; we therefore need to increase 191.8 by the same proportion.

$$\text{Thus } \frac{X}{191.8} = \frac{347.2}{100}$$

$$\text{Hence } X = \frac{191.8 \times 347.2}{100} = 665.9$$

Between January 1962 and September 1983, the index of retail prices increased from 100 to 665.9 – an increase by a factor approaching 7.

Exercise 4

This is a continuation of Exercise 3.
(i)   If the index of industrial production for all industries for 1980 was 106 with 1975 = 100, find the index of industrial production for all industries for August 1983 with 1975 = 100.
(ii) Write a brief report to the Managing Director of your findings.

(ICSA (part) Dec. 1984)

(*Answer:* 95.3)

C. SOLUTIONS TO EXERCISES

S1

p_o	q_o	p_n	q_n	$p_o q_o$	$p_n q_o$	$p_o q_n$	$p_n q_n$
1.2	4,000	2.5	2,000	4,800	10,000	2,400	5,000
5.3	1,000	5.8	800	5,300	5,800	4,240	4,640
2.3	2,000	2.7	4,000	4,600	5,400	9,200	10,800
Totals				14,700	21,200	15,840	20,440

Laspeyres price index

$$= \frac{\Sigma p_n q_0}{\Sigma p_0 q_0} \times 100 = \frac{21,200}{14,700} \times 100 = 144.2$$

Paasche price index

$$= \frac{\Sigma p_n q_n}{\Sigma p_0 q_n} \times 100 = \frac{20,440}{15,840} \times 100 = 129.0$$

The main reason for the difference is that much smaller quantities are now purchased of the item (plastic) which has had the largest increase in price.

S2

Using the table of solution S1 above:

Laspeyres quantity index

$$= \frac{\Sigma p_0 q_n}{\Sigma p_0 q_0} \times 100 = \frac{15,840}{14,700} \times 100 = 107.8$$

Paasche quantity index

$$= \frac{\Sigma p_n q_n}{\Sigma p_n q_0} \times 100 = \frac{20,440}{21,200} \times 100 = 96.4$$

S3

W	I	WI
264	101	26,664
25	82	2,050
41	88	3,608
68	95	6,460
325	84	27,300
99	92	9,108
52	82	4,264
126	83	10,458
1,000		89,912

(i) Index $= \frac{\Sigma WI}{\Sigma W} = \frac{89,912}{1,000} = 89.9$

(ii) Exclude first item:
$\Sigma WI = 89,912 - 26,664$
$\quad = 63,248$
$\Sigma W = 1,000 - 264 = 736$

Index $= \frac{\Sigma WI}{\Sigma W} = \frac{63,248}{736}$

$\quad = 85.9$

S4

(i)
$$
\begin{array}{lll}
1975 & 100 & \\
1980 & 106 & 100 \\
\text{August 1983} & X & 89.9
\end{array}
$$

$$\frac{X}{106} = \frac{89.9}{100} \qquad X = \frac{106 \times 89.9}{100} = 95.3$$

(ii) Between 1975 and August 1983 the index fell from 100 to 95.3, thus an increase of 24% of the company is an above average performance.

D. RECENT EXAMINATION QUESTIONS

Q1

Prodco plc manufactures an item of domestic equipment which requires a number of components which have varied as various modifications of the model have been produced.

Table Q1 shows the number of components required together with their price over the last three years of production.

Table Q1

Component	1981 Price (£)	1981 Quantity	1982 Price (£)	1982 Quantity	1983 Price (£)	1983 Quantity
A	3.63	3	4.00	2	4.49	2
B	2.11	4	3.10	5	3.26	6
C	10.03	1	10.36	1	12.05	1
D	4.01	7	5.23	6	5.21	5

(a) Establish the base weighted price indices for 1982 and 1983 based on 1981 for the item of equipment.

(b) Establish the current weighted price indices for 1982 and 1983 based on 1981 for the item of equipment.

(c) Using the results of (a) and (b) as illustrations compare and contrast Laspeyres and Paasche price index numbers.

(CACA (ACCA) Dec. 1984)

(*Answers:* 124.3, 130.6; 125.7; 133.3)

Q2

(a) Explain the function of index numbers and state why they are useful.

(b) From the data in Table Q2 calculate both the Laspeyres price index and Paasche price index.

Table Q2

Commodity	Unit of purchase	Price per unit (£) Base period	Price per unit (£) Period 1	Price per unit (£) Period 2
A	2 gallon drum	36	40	42
B	1 ton	80	90	100
C	10 pounds	45	41	41
D	1 kilogram	15	16	18
E	100 yards	5	6	6
F	1 cwt	150	150	180

Commodity	Quantity in units Base period	Quantity in units Period 1	Quantity in units Period 2
A	100	95	90
B	12	10	10
C	16	18	20
D	115	120	120
E	1,100	1,200	1,400
F	70	60	60

(ICMA May 1975)

(*Answers:* 107.3; 118.8; 107.9; 118.6)

Table Q3 shows the index of industrial production in the United Kingdom July 1982.

Table Q3

	Weight	Index (1975 = 100)
Mining and quarrying*	41	361
Manufacturing		
Food, drink and tobacco	77	106
Chemicals	66	109
Metal	47	72
Engineering	298	86
Textiles	67	70
Other manufacturing	142	91
Construction	182	84
gas, electricity and water	80	115

*Including North Sea oil

(a) Calculate the index of industrial production for (i) all industries; (ii) all industries except mining and quarrying; (iii) manufacturing industries.
(b) Comment on your results.

(ICSA June 1983)

(*Answers:* 101.3; 90.2; 88.9)

E. OUTLINE ANSWERS TO EXAM QUESTIONS

A1

(a)

p_0	q_0	p_2	p_3	$p_0 q_0$	$p_2 q_0$	$p_3 q_0$
3.63	3	4.00	4.49	10.89	12.00	13.47
2.11	4	3.10	3.26	8.44	12.40	13.04
10.03	1	10.36	12.05	10.03	10.36	12.05
4.01	7	5.23	5.21	28.07	36.61	36.47
Totals				57.43	71.37	75.03

Laspeyres price index 1982

$$= \frac{\Sigma p_2 q_0}{\Sigma p_0 q_0} \times 100 = \frac{71.37}{57.43} \times 100 = 124.3$$

Laspeyres price index 1983

$$=\frac{\Sigma p_3 q_0}{\Sigma p_0 q_0} \times 100 = \frac{75.03}{57.43} \times 100 = 130.6$$

(b)

p_0	p_2	q_2	p_3	q_3	$p_0 q_2$	$p_2 q_2$	$p_0 q_3$	$p_3 q_3$
3.63	4.00	2	4.49	2	7.26	8.00	7.26	8.98
2.11	3.10	5	3.26	6	10.55	15.50	12.66	19.56
10.03	10.36	1	12.05	1	10.03	10.36	10.03	12.05
4.01	5.23	6	5.21	5	24.06	31.38	20.05	26.05
Totals					51.90	65.24	50.00	66.64

Paasche price index 1982

$$=\frac{\Sigma p_2 q_2}{\Sigma p_0 q_2} \times 100 = \frac{65.24}{51.90} \times 100 = 125.7$$

Paasche price index 1983

$$=\frac{\Sigma p_3 q_3}{\Sigma p_0 q_3} \times 100 = \frac{66.64}{50.00} \times 100 = 133.3$$

(c) See Worked Example 1 in the text.

A2

p_0	q_0	p_1	p_2	$p_0 q_0$	$p_1 q_0$	$p_2 q_0$
36	100	40	42	3,600	4,000	4,200
80	12	90	100	960	1,080	1,200
45	16	41	41	720	656	656
15	115	16	18	1,725	1,840	2,070
5	1,100	6	6	5,500	6,600	6,600
150	70	150	180	10,500	10,500	12,600
Totals				23,005	24,676	27,326

Laspeyres price index (1)

$$=\frac{\Sigma p_1 q_0}{\Sigma p_0 q_0} \times 100 = \frac{24,676}{23,005} \times 100 = 107.3$$

Laspeyres price index (2)

$$=\frac{\Sigma p_2 q_0}{\Sigma p_0 q_0} \times 100 = \frac{27,326}{23,005} \times 100 = 118.8$$

p_0	p_1	q_1	p_2	q_2	p_0q_1	p_1q_1	p_0q_2	p_2q_2
36	40	95	42	90	3,420	3,800	3,240	3,780
80	90	10	100	10	800	900	800	1,000
45	41	18	41	20	810	738	900	820
15	16	120	18	120	1,800	1,920	1,800	2,160
5	6	1,200	6	1,400	6,000	7,200	7,000	8,400
150	150	60	180	60	9,000	9,000	9,000	10,800
Totals					21,830	23,558	22,740	26,960

Paasche price index (1)

$$= \frac{\Sigma p_1 q_1}{\Sigma p_0 q_1} \times 100 = \frac{23,558}{21,830} \times 100 = 107.9$$

Paasche price index (2)

$$= \frac{\Sigma p_2 q_2}{\Sigma p_0 q_2} \times 100 = \frac{26,960}{22,740} \times 100 = 118.6$$

A3

W	I	WI
41	361	14,801
77	106	8,162
66	109	7,194
47	72	3,384
298	86	25,628
67	70	4,690
142	91	12,922
182	84	15,288
80	115	9,200
1,000		101,269

(i) Index $= \dfrac{\Sigma WI}{\Sigma W} = \dfrac{101,269}{1,000} = 101.3$

(ii) Exclude first item:
$\Sigma WI = 101,269 - 14,801 = 86,468$
$\Sigma W = 1,000 - 41 = 959$

Index $= \dfrac{\Sigma WI}{\Sigma W} = \dfrac{86,468}{959} = 90.2$

(iii) Exclude first and last two items:
$\Sigma WI = 101,269 - 14,801 - 15,288 - 9,200 = 61,980$
$\Sigma W = 1,000 - 41 - 182 - 80 = 697$

Index $= \dfrac{\Sigma WI}{\Sigma W} = \dfrac{61,980}{697} = 88.9$

A STEP FURTHER   Mulholland and Jones, *Fundamentals of Statistics*, Ch. 15.

Chapter 8

Time series and forecasting

A. GETTING STARTED

A *time series* is a set of values which occur sequentially in time. The daily takings of a shop would produce a time series. Many economic and business indicators such as the index of retail prices, unemployment statistics, the index of industrial production, etc., form a time series.

Examination questions on time series are set frequently. There are two types of question: the first, and most popular, involves *quarterly* data; the second involves *daily* or *annual* data. It is desirable to know how to handle both types of data, but if time is short, concentrate on Worked Example 2 which involves quarterly data.

A time series consists of four components:

(i)   *Trend:* the underlying movement of the series.
(ii) *Cyclical:* the movement of a series due to booms and slumps.
(iii) *Seasonal factor:* the movement of a series due to the effects of changes in season. For *quarterly* data, this involves the effect of spring, summer, autumn and winter on many economic and business series; for *daily* data, it may involve the effect of the day of the week on (say) sales.
(iv) *Residual or random factor:* some factors which affect a time series occur at random. Unpredictable factors such as strikes, exceptionally bad weather, sudden shortage of materials, etc., are examples.

Examination questions usually involve less than 20 observations. It is *not* possible to find cyclical movements with so few observations. *In an elementary treatment of time series the cyclical and trend components are combined and are regarded as the trend.*

B. ESSENTIAL PRINCIPLES

ADDITIVE MODEL

Most examination questions involve the *additive model*. This model assumes that an observation (Y) in a time series is the result of *adding* algebraically the trend (T), the seasonal factor (S) and the residual factor (R). This may be expressed as follows:

$$Y = T + S + R$$

MULTIPLICATIVE MODEL

The additive model does, in effect, assume that the seasonal and random factors are independent of the trend. In the case of the unemployment statistics in the United Kingdom, the seasonal factors which affect unemployment (extra jobs which are available in the summer and at Christmas reduce unemployment at those times) are found to be unrelated to the level of unemployment. Thus for the *unemployment* statistics the *additive* model is appropriate. In the case of production, however, if the level of production increases, the seasonal factors (many factories close for a period in July or August for holidays) tend to operate *in proportion to* the level of production. When the seasonal or residual factors *are* proportional to the trend, as in the case of *production*, the *multiplicative model* should be used.

The multiplicative model assumes that an observation (Y) is the *product* of the trend (T), the seasonal factor (S) and the residual factor (R). This may be expressed as follows:

$$Y = T \times S \times R$$

CALCULATION OF THE TREND

It is possible to find a straight-line *trend* by the use of a regression line (see Chapter 6). The standard method of finding the trend is, however, the *method of moving averages*. Examination questions usually specify which method you have to employ.

If you are finding a moving average for an *even* number of terms it is necessary to find a *centred* moving average. How to do this is explained in Worked Example 2 using quarterly data. The number of terms to use in a moving average depends on the *period* of the data, i.e. the time between successive peaks or troughs in the data. If you have *annual* data and the time between successive booms or slumps is about 7 years, then the period would be 7. If you have *quarterly* data, then the period would be 4. If you have *daily* data, based on, say, takings during a 5-day week, then the period would be 5.

Worked Example 1

The data in Table 8.1 gives the takings of a shop during the last three weeks.

Table 8.1

	Monday	Tuesday	Wednesday	Thursday	Friday
Week 1	128	168	80	190	230
Week 2	142	186	84	201	240
Week 3	150	194	86	210	262

(i) By means of a moving average find the trend; (ii) obtain the average daily variations; (iii) calculate the residuals; (iv) adjust the takings for week 3 for daily variation.

We shall use a 5-term moving average since the period of the daily data is 5. The first step is to find the moving total, so add the first 5 terms $(128 + 168 + 80 + 190 + 230 = 796)$. Next, add the second to the sixth terms $(168 + 80 + 190 + 230 + 142 = 810)$, and continue in this way. The results are given in column (ii) of Table 8.2.

Table 8.2

Takings	5-term moving total	5-term moving average	col. (i)– col. (iii)	Adjusted takings	Residual variation
col. (i)	col. (ii)	col. (iii)	col. (iv)	col. (v)	col. (vi)
128				152	
168				150	
80	796	159.2	-79.2	167	8
190	810	162	28.0	162	0
230	828	165.6	64.4	165	-1
142	832	166.4	-24.4	166	0
186	843	168.6	17.4	168	-1
84	853	170.6	-86.6	171	0
201	861	172.2	28.8	173	1
240	869	173.8	66.2	175	1
150	871	174.2	-24.2	174	0
194	880	176.0	18.0	176	0
86	902	180.4	-94.4	173	-7
210				182	
262				197	

In column (ii) it will be noted that the first moving total is placed at the third entry, i.e. the middle entry of the 5 terms totalled. To find the trend we calculate the *moving average* – i.e. we divide the moving total by 5. The moving average is entered in col. (iii), and this represents the *trend* of the data.

SEASONAL VARIATIONS

The *difference* between the trend, col. (iii), and the data, col. (i), is made up of *seasonal* and *residual* factors. In Table 8.2 this difference is found in col. (iv). To find the average *seasonal factor* (in Worked Example 1 the 'seasonal' factor is the *daily* variation), we take the seasonal factor for each day and average. The very process of *averaging* removes the random or residual factor (R), leaving us with the seasonal factor (S) we seek. The easiest way to achieve this is to enter the data from col. (iv) as in Table 8.3.

Table 8.3

	Monday	Tuesday	Wednesday	Thursday	Friday
Week 1	—	—	−79.2	28.0	64.4
Week 2	−24.4	17.4	−86.6	28.8	66.2
Week 3	−24.2	18.0	−94.4	—	—
Total	−48.6	35.4	−260.2	56.8	130.6
Average	−24.3	17.7	−86.7	28.4	65.3

As the original data was given to the nearest integer, the seasonal factors should be rounded to the nearest integer. The results are given below:

Monday	Tuesday	Wednesday	Thursday	Friday
−24	18	−87	28	65

The seasonal adjustments should sum to zero. In this case they happen to sum to zero. When they do not sum to zero (this can occur because of a residual factor), some adjustment is necessary. If the seasonal factors had summed to (say) +5, then 1 could be subtracted from each seasonal factor. If the sum is not a multiple of 5, say 3, then only three of the factors would need to be reduced by 1.

SEASONALLY ADJUSTED DATA

In column (v) of Table 8.2 the takings have been *seasonally adjusted*. The seasonal adjustment is obtained by subtracting algebraically the seasonal factors from the data. Thus for the first observation (Monday week 1) 128, we subtract −24, i.e. $128 - (-24) = 128 + 24 = 152$; for the second observation $168 - 18 = 150$; and so on.

RESIDUAL FACTOR

On page 111 it was explained that an observation is the algebraic sum of the trend, seasonal and residual factors. In col. (iv) of Table 8.2 we have subtracted the trend from the data, so we are left with seasonal and residual factors. If we now *subtract* algebraically *the seasonal factors* found above from col. (iv) we are left with the *residual* factor. For the first value in col. (iv) (Wednesday week 1) −79.2, we subtract −87, i.e. $-79.2 - (-87) = -79.2 + 87 = 7.8 = 8$ to the nearest integer; for the second value $28.0 - 28 = 0$; and so on.

From the data supplied in Table 8.4 calculate the trend, the seasonal or, in this case, the average daily variations, and the residuals.

The figures represent cash sales of a store which does not open for business on a Monday.

Table 8.4

Week number: Day of week	1 (£)	2 (£)	3 (£)	4 (£)
Tuesday	360	350	380	390
Wednesday	400	430	440	450
Thursday	480	490	490	500
Friday	600	580	590	600
Saturday	660	680	690	690

(ICMA May 1979)

Note: For this chapter see solutions for answers.

Worked Example 2

Table 8.5 shows United Kingdom passenger movement abroad by sea and air (figures in 100,000 passengers).

Table 8.5

Quarter	1	2	3	4
1979	46	86	120	61
1980	53	89	125	61
1981	51	91	132	66

(i) By means of a moving average find the trend and the seasonal adjustments.
(ii) Forecast passenger movements for the first two quarters of 1982.

(ICSA June 1983)

QUARTERLY DATA

When we have quarterly data, as in worked example 2, the period is 4, but summing 4 observations and dividing by 4 does not provide a suitable trend for finding seasonal factors. It is necessary to find a *centred* moving average. The simplest procedure for finding a centred moving average is set out in Table 8.6.

Table 8.6

Year and quarter		Data	Add in fours	Add col. (ii) in twos	Trend: divide col. (iii) by 8	col. (i) − col. (iv)
		col. (i)	col. (ii)	col. (iii)	col. (iv)	col. (v)
1979	1	46				
	2	86				
			313			
	3	120		633	79.125	40.875
			320			
	4	61		643	80.375	− 19.375
			323			
1980	1	53		651	81.375	− 28.375
			328			
	2	89		656	82	7
			328			
	3	125		654	81.75	43.25
			326			
	4	61		654	81.75	− 20.75
			328			
1981	1	51		663	82.875	− 31.875
			335			
	2	91		675	84.375	6.625
			340			
	3	132				
	4	66				

The *centred moving average* is given in col. (iv) and can be used for finding the seasonal factors. If we had found the moving average by dividing col. (ii) by 4, the moving average would have been placed *in between two quarters* and would not have been of any use in finding seasonal factors.

The next step is to find the average seasonal factors, as in Table 8.7.

Table 8.7

Quarter	1	2	3	4
1979	—	—	40.875	− 19.375
1980	− 28.375	7	43.25	− 20.75
1981	− 31.875	6.625	—	—
Totals	− 60.25	13.625	84.125	− 40.125
Average	− 30.125	6.8125	42.0625	− 20.0625
Seasonal factors	− 30	7	42	− 20

The seasonal factors are given to the same accuracy as the original data. The seasonal factors sum to − 1; an adjustment is therefore needed. Add 1 to the largest value so that the seasonal factors become:

$$-30 \qquad 7 \qquad 43 \qquad -20$$

FORECASTING

Most forecasting procedures involve projecting the trend forward. This can be done in various ways; for example, it is possible to plot the trend and to project this forward. In an examination you will not have much time. The quickest procedure to use in an examination is to take the last trend value (84.375) *and* to subtract the first trend value (79.125), which gives + 5.25. As the trend in this example is upwards we obtain a positive value. There are 7 increments between the first and last trend value, thus the *average* increment is + 5.25 divided by 7, i.e. 0.75. To find the trend values for the first quarter of 1982, take the last trend value 84.375 (which relates to the second quarter of 1981) and add on 3 times the increment (0.75). Finally, to obtain the *forecast* add on algebraically the seasonal factor. The results are set out below:

Forecast 1982 quarter 1: $84.375 + 3 \times 0.75 + (-30) = 56.625 = 57$
Forecast 1982 quarter 2: $84.375 + 4 \times 0.75 + (+7) = 94.375 = 94$

This procedure assumes that there is a *linear* trend and that the trend will *continue to be linear*.

Exercise 2

Table 8.8 shows quarterly sales by value (£1,000) for 1971–74.

Table 8.8

| Year | Quarters | | | |
	1	2	3	4
1971	216	335	349	297
1972	243	376	368	305
1973	270	385	362	314
1974	300	398	375	326

Using the method of moving averages, find the trend and the average seasonal factors. Thus estimate the sales figures for the first two quarters of 1975.

(ICSA Dec. 1976)

MULTIPLICATIVE MODEL

Worked Examples 1 and 2 assumed an *additive* model. Most examination questions do not specify which model to use, so you can omit this section if you have limited time.

Worked Example 3

Using the data of Worked Example 2 find the trend and seasonal factors, and forecast passenger movements for the first two quarters of 1982.

The calculation of the trend (Table 8.9) is the same as for the additive model, so columns (i) to (iv) are the same as in Table 8.8.

Table 8.9

Year and quarter		Data col. (i)	Add in fours col. (ii)	Add col. (ii) in twos col. (iii)	Trend: divide col. (iii) by 8 col. (iv)	col. (i)/ col. (iv) × 100 col. (v)
1979	1	46				
	2	86				
	3	120	313	633	79.125	151.7
	4	61	320	643	80.375	75.9
1980	1	53	323	651	81.375	65.1
	2	89	328	656	82	108.5
	3	125	328	654	81.75	152.9
	4	61	326	654	81.75	74.6
1981	1	51	328	663	82.875	61.5
	2	91	335	675	84.375	107.9
	3	132	340			
	4	66				

To find the seasonal factors using the multiplicative model, *divide* col. (i) by col. (iv) and then multiply by 100. In other words, the original data $Y(= T \times S \times R)$, divided by T, gives $S \times R$, i.e. the seasonal factors including a random element. Multiplying by 100 expresses $S \times R$ as a percentage. *Averaging* for each quarter, as in Table 8.10, then removes the random or residual factor, R, leaving us with the seasonal factor, S.

Table 8.10

Quarter	1	2	3	4
1979	—	—	151.7	75.9
1980	65.1	108.5	152.9	74.6
1981	61.5	107.9	—	—
Totals	126.6	216.4	304.6	150.5
Average	63.3	108.2	152.3	75.3

The sum of the seasonal factors should be 400. In fact the total is 399.1 – a shortfall of 0.9. As a result 0.2 is added to three of the quarters and 0.3 to the largest quarter to produce the following seasonal factors:

63.5	108.4	152.6	75.5

The method of forecasting future trend values is the same as for the additive model. For the actual forecasts we take percentages of the trend values.

Forecast 1982 quarter 1: $(84.375 + 3 \times 0.75) \times 63.5 \div 100 = 55$
Forecast 1982 quarter 2: $(84.375 + 4 \times 0.75) \times 108.4 \div 100 = 95$

Exercise 3

Using the data of Exercise 2 and employing the multiplicative model, find the trend and seasonal factors, and forecast sales for the first two quarters of 1975.

C. SOLUTIONS TO EXERCISES

S1

(i)	(ii)	(iii)	(iv)	(i)	(ii)	(iii)	(iv)
360				380	2,570	514	−134
400				440	2,580	516	−76
480	2,500	500	−20	490	2,590	518	−28
600	2,490	498	102	590	2,600	520	70
660	2,520	504	156	690	2,610	522	168
350	2,530	506	−156	390	2,620	524	−134
430	2,510	502	−72	450	2,630	526	−76
490	2,530	506	−16	500	2,630	526	−26
580	2,560	512	68	600			
680	2,570	514	166	690			

The trend is given in column (iii).

Week	Tuesday	Wednesday	Thursday	Friday	Saturday
1	—	—	−20	102	156
2	−156	−72	−16	68	166
3	−134	−76	−28	70	168
4	−134	−76	−26	—	—
Totals	−424	−224	−90	240	490
Averages	−141.33	−74.67	−22.5	80	163.33

The seasonal factors add to $+4.83$; subtract 0.97 from each and round. Thus the seasonal factors are $-142, -76, -23, 79, 162$.

The residuals are found by subtracting seasonal factors from column (iv):

$$3, 23, -6, -14, 4, 7, -11, 4, 8, 0, -5, -9, 6, 8, 0, -3$$

S2

Note: Column (vi) below is required for solution S3.

(i)	(ii)	(iii)	(iv)	(v)	(vi)
216					
335					
349	1,197	2,421	302.625	46.375	115.3
297	1,224	2,489	311.125	−14.125	95.5
243	1,265	2,549	318.625	−75.625	76.3
376	1,284	2,576	322	54	116.8
368	1,292	2,611	326.375	41.625	112.8
305	1,319	2,647	330.875	−25.875	92.2
270	1,328	2,650	331.25	−61.25	81.5
385	1,322	2,653	331.625	53.375	116.1
362	1,331	2,692	336.5	25.5	107.6
314	1,361	2,735	341.875	−27.875	91.8
300	1,374	2,761	345.125	−45.125	86.9
398	1,387	2,786	348.25	49.75	114.3
375	1,399				
326					

Year	I	II	III	IV
1971	—	—	46.375	−14.125
1972	−75.625	54	41.625	−25.875
1973	−61.25	53.375	25.5	−27.875
1974	−45.125	49.75	—	—
Totals	−182	157.125	113.5	−67.875
Averages	−60.67	52.38	37.83	−22.63

Seasonal factors add to $+6.91$; subtract 1.73 from each and round. Thus the seasonal factors are $-62, 51, 36, -24$. As these still add to $+1$ subtract 1 from first quarter $= -63$.

To forecast, find quarterly trend increment $348.25 - 302.625 = 45.625$, and divide by 11 $= 4.15$.

Forecast quarter $1 = 348.25 + 3 \times 4.15 - 63 = 298$
Forecast quarter $2 = 348.25 + 4 \times 4.15 + 51 = 416$

S3

The calculations for trend is given in solution S2.

Year	I	II	III	IV
1971	—	—	115.3	95.5
1972	76.3	116.8	112.8	92.2
1973	81.5	116.1	107.6	91.8
1974	86.9	114.3	—	—
Totals	244.7	347.2	335.7	279.5
Averages	81.6	115.7	111.9	93.2

The seasonal factors add to 402.4, subtract 0.6 from each quarter.
Seasonal factors are thus 81.0, 115.1, 111.3, 92.6.
Using calculation from solution S2:

$$\text{Forecast quarter } 1 = (348.25 + 3 \times 4.15) \times 81 \div 100 = 292$$
$$\text{Forecast quarter } 2 = (348.25 + 4 \times 4.15) \times 115.1 \div 100 = 420$$

D. RECENT EXAMINATION QUESTIONS

Q1

Table Q1

Quarter	1	2	3	4
1979	486	372	321	466
1980	479	315	283	441
1981	448	308	291	477

Table Q1 shows production of cattle food (thousands tonnes).

(a) By means of a centred moving average calculate the trend and seasonal variations.

(b) Suggest why these data have this pattern of seasonal variations. Would you expect a similar pattern for the production of poultry food? Give a reason for your answer.

(RSA June 1984)

Q2

Table Q2

Period	1976	1977	1978	1979
Jan.–Apr.	84	92	94	100
May–Aug.	121	140	148	161
Sept.–Dec.	67	81	83	95

A company finds that the sales of its products are affected by the time of year, which tends to obscure any trend. The sales data in Table Q2 shows sales in successive periods of 4 months.

(a) Using the method of moving averages, estimate the trend in these data and calculate the 'seasonal' effects.

(b) Use your trend found in part (a) to estimate the mean monthly increase in sales that the company has achieved over the period.

(c) How much more would the company expect to sell in May–Aug. than in Sept.–Dec. in a situation when there is no trend?

(CACA (ACCA) Dec. 1981)

Q3

Table Q3

Quarters	I	II	III	IV
1979	77.5	63.2	54.8	78.5
1980	80.2	66.0	59.3	84.7
1981	83.8	68.4	62.1	90.3

Table Q3 shows sales of microcomputers by value (£'000).

Using a method of moving averages, find the trend from the data and, after extracting the seasonal deviations (by the use of the additive method), forecast the sales by value for the first two quarters of 1982.

(LCCI May 1983)

Q4

Table Q4

Quarter	I	II	III	IV
1977	16.2	12.4	7.6	17.8
1978	18.2	14.0	8.3	19.5
1979	17.6	13.4	8.4	18.2
1980	16.5	12.4		

Table Q4 shows sales of colour television sets (sales in thousands).

Extract the trend from the data using the method of moving averages. Show the original data and the trend superimposed on a historigram. Hence estimate the sales for the region for the last two quarters of 1980.

(ICSA Dec. 1980)

E. OUTLINE ANSWERS TO EXAM QUESTIONS

A1

486				
372				
321	1,645	3,283	410.375	− 89.375
466	1,638	3,219	402.375	63.625
479	1,581	3,124	390.5	88.5
315	1,543	3,061	382.625	− 67.625
283	1,518	3,005	375.625	− 92.625
441	1,487	2,967	370.875	70.125
448	1,480	2,968	371	77
308	1,488	3,012	376.5	− 68.5
291	1,524			
477				

Year	I	II	III	IV
1979	—	—	− 89.375	63.625
1980	88.5	− 67.625	− 92.625	70.125
1981	77	− 68.5	—	—
Totals	165.5	− 136.125	− 182	133.75
Averages	82.75	− 68.06	− 91	66.88

121

These add to -9.43; add 2.36 to each quarter and round.
Thus seasonal factors are 85, -66, -89, 69.
These add to -1; add 1 to third quarter $= -88$.

In United Kingdom cattle graze in summer months, therefore less cattle food needed. Most poultry are kept in batteries and are fed throughout year, thus no seasonality is to be expected.

A2

(a)

84				94	323	107.67	-13.67
121	272	90.67	30.33	148	325	108.33	39.67
67	280	93.33	-26.33	83	331	110.33	-27.33
92	299	99.67	-7.67	100	344	114.67	-14.67
140	313	104.33	35.67	161	356	118.67	42.33
81	315	105	-24	95			

Year	Jan.–Apr.	May–Aug.	Sept.–Dec.
1976	—	30.33	-26.33
1977	-7.67	35.67	-24
1978	-13.67	39.67	-27.33
1979	-14.67	42.33	—
Totals	-36	148	-77.67
Averages	-12	36	-25.89

The seasonal factors add to -1.89; add 1 each to largest and round. Seasonal factors are -12, 37, -25.

(b) Increase in trend $= 118.67 - 90.67 = 28$. Over 4-month periods this represents an increase of $28 \div 9 = 3.11$. Thus mean monthly increase is one-quarter of $3.11 = 0.78$.

(c) $37 - (-25) = 62$.

A3

77.5				
63.2	274			
54.8	276.7	550.7	68.84	-14.04
78.5	279.5	556.2	69.53	8.97
80.2	284	563.5	70.44	9.76
66.0	290.2	574.2	71.78	-5.78
59.3	293.8	584	73	-13.7
84.7	296.2	590	73.75	10.95
83.8	299	595.2	74.4	9.4
68.4	304.6	603.6	75.45	-7.05
62.1				
90.3				

Year	I	II	III	IV
1979	—	—	− 14.04	8.97
1980	9.76	− 5.78	− 13.7	10.95
1981	9.4	− 7.05	—	—
Totals	19.16	− 12.83	− 27.74	19.92
Averages	9.58	− 6.42	− 13.87	9.96

Seasonal factors add to -0.75; add 0.19 and round.
Seasonal factors are then 9.8, -6.2, -13.7, 10.1.
Trend increase $= 75.45 - 68.84 = 6.61$; divide by $7 = 0.94$.
Forecast quarter $1 = 75.45 + 3 \times 0.94 + 9.8 = 88.1$
Forecast quarter $2 = 75.45 + 4 \times 0.94 - 6.2 = 73$

A4

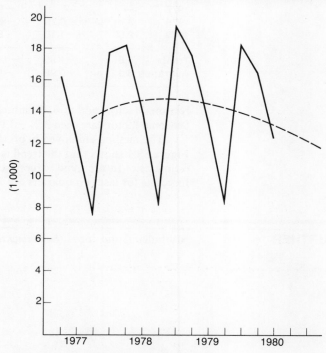

Fig. 8S.1   Sales of colour television sets

16.2				
12.4				
7.6	54	110	13.75	−6.15
17.8	56	113.6	14.2	3.6
18.2	57.6	115.9	14.49	3.71
14	58.3	118.3	14.79	−0.79
8.3	60	119.4	14.93	−6.63
19.5	59.4	118.2	14.78	4.72
17.6	58.8	117.7	14.71	2.89
13.4	58.9	116.5	14.56	−1.16
8.4	57.6	114.1	14.26	−5.86
18.2	56.5	112	14	4.2
16.5	55.5			
12.4				

Year	I	II	III	IV
1977	—	—	−6.15	3.6
1978	3.71	−0.79	−6.63	4.73
1979	2.89	−1.16	−5.86	4.2
Totals	6.6	−1.95	−18.64	12.53
Averages	3.3	−0.98	−6.21	4.18

Seasonal factors add to 0.29; subtract 0.07 from each and round. Seasonal factors are then 3.2, −1.0, −6.3, 4.1.

The method employed in previous questions cannot be used here. Figure 8S.1 shows that the trend is a curve. Projecting the trend, estimates of future trend values are 12.7 and 12.3 respectively. Thus forecasts for last two quarters are $12.7 - 6.3 = 6.4$ and $12.3 + 4.1 = 16.4$.

A STEP FURTHER

Mulholland and Jones, *Fundamentals of Statistics*, Ch. 16.

Chapter 9   # Probability

A. GETTING STARTED

You should first check whether probability is in your syllabus – at the time of writing probability is not included in the CACA (ACCA) syllabus for numerical analysis and data processing, but probability *is* included in most foundation-level syllabuses and in many diploma and undergraduate courses. You should check from past examination papers whether questions on probability are regularly set. Most examination candidates avoid probability questions, and those that attempt probability questions often do them very badly. If probability questions are regularly set and there is a limited choice of questions you will have to become reasonably proficient at answering probability questions. If probability questions are set occasionally and you have a good choice of questions then you could avoid answering probability questions in an examination.

The typical examination answer to probability questions is either almost completely correct or almost completely wrong. If the answer is correct then high marks are scored quickly, so if you are good at probability (it is likely that you met probability in 'O'-level mathematics so you will have some idea of how good you are), then such a question is worth trying, otherwise probability is best avoided. If you plan to avoid answering probability questions in the examination room, you should, nevertheless, study this chapter (perhaps omitting the section on Recent Examination Questions), simply because *some* knowledge of probability is essential for Chapters 10 and 11.

Examination candidates are usually more successful in obtaining the correct answer if the question involves a diagram – a tree or a Venn diagram. When there is a choice of method of solution, a diagrammatic approach is recommended.

B. ESSENTIAL PRINCIPLES

CLASSICAL PROBABILITY

Problems where you have to work out the probability of the outcome, as when tossing a coin, drawing a card from a pack of playing cards or rolling a true dice, are often referred to as *classical probability*. Most 'O'-level mathematics questions on this topic involve classical probability.

STATISTICAL PROBABILITY

A manufacturer might wish to find the probability that he will obtain a given percentage of satisfactory parts from the *mass production* of such parts. This is a problem involving *statistical probability*, and is considered in detail in Chapter 10.

SUBJECTIVE PROBABILITY

A marketing manager might wish to estimate the probability that the launch of a new product will be 'successful'. He will take into account past experience of the launch of similar products; the likely state of the economy next year, etc. His estimate of the probability will depend on his subjective judgement.

This chapter will focus on the problems of *classical* probability.

RULES OF PROBABILITY

If a true six-sided dice is rolled, let A be the event a six appears then $P(A) = 1/6$. If $\bar{A}$ is the event that a six does *not* appear, then $P(\bar{A}) = 5/6$. Note that $P(A) + P(\bar{A}) = 1$ and $P(\bar{A}) = 1 - P(A)$.

Independent events

Two or more events are said to be *independent* if the outcome of one event makes no difference to the outcome of the other event(s). In examination questions it is essential to determine whether events are dependent or independent. To illustrate: suppose a bag contains 4 red and 6 blue balls. If we draw a ball at random from the bag, the probability of a ball being red is $4/10 = 0.4$. If we draw a second ball from the bag, what is the probability that the second ball is red? If we replace the first ball into the bag *before* the second draw, the probability that the ball is red is 0.4 and is **independent** of the first draw. However, if the first ball (which we are assuming is red) is *not* replaced, then there are 9 balls in the bag and the number of red balls is 3. The probability of drawing a red ball is then $3/9 = 1/3$ and is **dependent** on the outcome of the first draw. The key point is whether the question states 'with replacement' or 'without replacement'.

Mutually exclusive events

Suppose there are several possible outcomes of an event. The events are said to be *mutually exclusive* if it is impossible for more than one outcome to occur at any one time.

Consider the price of a share at the end of a month compared with the price at the beginning of the month. Let the possible outcomes be: A, a rise in price of more than 10%; B, a rise in price of less than 10%; C, no change; and D, a fall in price. Clearly only one outcome is possible, i.e. events A, B, C and D are mutually exclusive events.

ADDITION LAW OF PROBABILITIES

If events are mutually exclusive, then the probability that two or more outcomes will occur is the *sum* of the probabilities. In symbols, this may be expressed as

$$P(A \text{ or } B) = P(A) + P(B)$$

or generally

$$P(A \text{ or } B \text{ or } C \ldots \text{ or } N) = P(A) + P(B) + P(C) + \ldots + P(N).$$

Unfortunately there is no uniform symbol for $P(A \text{ or } B)$, which means the probability of A or B. In *some* examination questions and text books this is expressed as $P(A \cup B)$ or, in general, $P(A \cup B \cup C \cup \ldots \cup N)$.

Worked Example 1

Consider the example on share prices discussed in the paragraph on mutually exclusive events. If $P(A) = 0.3$, $P(B) = 0.4$, $P(C) = 0.1$ and $P(D) = 0.2$, what is the probability that the share price will rise?

We need to find the probability of **either A or B** occurring.
$P(A \text{ or } B) = P(A) + P(B)$ $P(A \text{ or } B) = 0.3 + 0.4 = 0.7$

Exercise 1

A company tenders for two contracts A and B. The probability that it will obtain contract A is 0.2 and contract B is 0.3. What is the probability that it will obtain either contract A or contract B? (*Answer:* 0.5)

MULTIPLICATION LAW

If events A and B are *independent*, then the probability that events A and B will **both** occur is given by the *multiplication law*:

$$P(AB) = P(A) \times P(B)$$

or generally,

$$P(ABC \ldots N) = P(A) \times P(B) \times P(C) \times \ldots \times P(N)$$

Sometimes $P(AB)$ is expressed $P(A \cap B)$.

CONDITIONAL PROBABILITY

Let us return to the problem of the bag which contained 4 red balls and 6 blue balls. If we draw out two balls *without* replacement, what is the probability that both are red? Let A be the event 'first ball red' and B the event 'second ball red'. $P(A)$ is equal to 4/10. Now the probability of event B is *conditional* on the result of event A. If the first ball is red there are 3 red balls left out of the 9 in the bag, thus P (second ball red given first ball red) is 3/9. The probability is usually written $P(B/A)$, i.e. the probability of event B *given* that event A has occurred. The probability that both balls are red is equal to

$$\frac{4}{10} \times \frac{3}{9} = \frac{2}{15}$$

The multiplication law when events are *dependent* is

$$P(AB) = P(A) \times P(B/A)$$

Worked Example 2

A bag contains 20 balls of which 8 are red and 12 yellow. Three balls are drawn at random from the bag; what is the probability that they are all red if drawing is: (i) with replacement; (ii) without replacement?

(i) Here the events are *independent*, i.e. $P(ABC) = P(A) \times P(B) \times P(C)$
$P(A) = P(B) = P(C) = 8/20 = 0.4$ $P(ABC) = 0.4 \times 0.4 \times 0.4$
$= 0.064$

(ii) Here the events are *dependent*, i.e. $P(ABC) = P(A) \times P(B/A) \times P(C/AB)$.
From part (i) $P(A) = 0.4$

$$P(B/A) = \frac{7}{19} \qquad P(C/AB) = \frac{6}{18}$$

$$P(ABC) = \frac{8}{20} \times \frac{7}{19} \times \frac{6}{18} = \frac{14}{285} = 0.0491$$

Exercise 2

A box of 16 components contains 4 defective components. If 3 components are drawn from the box, what is the probability that they are all good: (i) if there is replacement; (ii) if there is no replacement?

If components are taken from the box one at a time without replacement and tested, what is the probability that the third component is the first good component?
(*Answers:* 0.4219; 0.3929; 0.04286)

Worked Example 3

A part is made up of five independently produced components, and the probability of a component being defective is 0.02. The *part* is defective if at least one *component* is defective. Find the probability that the part is defective.

The method here is to find the probability that the part is good and subtract this probability from 1.
Probability of a good component is $1 - 0.02 = 0.98$. The probability of 5 good components is $0.98^5 = 0.9039$, hence probability of a defective part is $1 - 0.9039 = 0.0961$.

Exercise 3

A manufacturer assembles a toy from four independently produced components, each of which has a probability of 0.01 of being defective. What is the probability of the toy being defective?

(ICMA (part) Nov. 1981)

(*Answer:* 0.0394)

TREE DIAGRAMS

Many examination questions are best solved using a *tree diagram*. The method is illustrated in Worked Example 4.

Three similar machines A, B and C are used to make a component. Machine A is new and produces 40% of the total output; machines B and C each produce 30% of the output. The percentage of defective components produced by machine A is 1%, and the corresponding percentages for machines B and C are 4% and 7% respectively.

(i) If a component is selected at random, find the probability that the component is defective.

(ii) If a component is found to be defective, find the probability that the component was made by machine C.

The first step is to draw the three branches of the tree corresponding to the machines A, B and C. These are labelled and the probabilities are written against each branch. For branch A add two further branches, one for defective components (labelled D) with a probability of 0.01 and one for good components (labelled G) with probability 0.99. This process should be repeated for branches B and C.

The probabilities along the branches should now be multiplied. The necessary multiplications are shown in Fig. 9.1; there are six multiplications and the total of these must be equal to 1, since this covers all eventualities.

We shall now use the tree to find the answers to the problem.

(i) The probability that the component is defective is found by adding up all the probabilities which involve D

i.e. $P(D) = 0.004 + 0.012 + 0.021$; thus $P(D) = 0.037$.

(ii) This is a *conditional* probability. We require $P(C/D)$, and from the section on conditional probability we use the result

$$P(DC) = P(D) \times P(C/D).$$

This may be rearranged

$$P(C/D) = \frac{P(DC)}{P(D)}$$

We have already found $P(D)$ in part (a). $P(DC)$ is the probability of a defective component from machine C and can be read from the tree $= 0.021$. Thus

$$P(C/D) = \frac{0.021}{0.037} = 0.5676$$

Part (ii) of this question involves the use of *Bayesian methods*. To solve problems it is not necessary to remember the formula used above – you can use the tree. You first have to find the probability of a defective by adding up the relevant branches of the tree. The next step is to find the relevant probability associated with machine C, then divide the smaller probability by the larger probability.

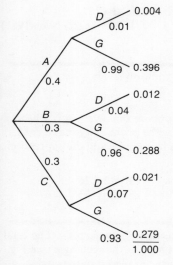

Fig. 9.1   Tree diagram

Exercise 4

A factory has a machine shop in which three machines (A, B and C) produce aluminium tubes. An inspector is equally likely to sample tubes from A and B, and three times as likely to select tubes from machine C as he is from B. The defective rates from these machines are:

A 10% B 10% C 20%

What is the probability that a tube selected by the inspector:
(i)   is from machine A;
(ii) is defective;
(iii) comes from machine A, given that it is defective?

(ICMA (part) Nov. 1981)

(*Answers:* 0.2; 0.16; 0.125)

ADDITION LAW OF PROBABILITIES WHEN EVENTS ARE MUTUALLY EXCLUSIVE

When events are *mutually exclusive* the addition law is:

$$P(A \text{ or } B) = P(A) + P(B)$$

If we have a pack of 52 playing cards, suppose that event A is a heart and event B is a club, then $P(A \text{ or } B)$, i.e. the probability that a card drawn at random is a heart or club, can be found using the expression above:

$$P(A) = \frac{13}{52} \qquad P(B) = \frac{13}{52} \qquad P(A \text{ or } B) = \frac{13}{52} + \frac{13}{52} = \frac{26}{52} = \frac{1}{2}$$

VENN DIAGRAM – ADDITION LAW OF PROBABILITIES WHEN EVENTS ARE NOT MUTUALLY EXCLUSIVE

This type of problem is best dealt with using a *Venn diagram*. The total sample is 52, which is represented by the area of the rectangle (see Fig. 9.2). The thirteen clubs and thirteen hearts are placed in the circles. The required probability is found by adding up the contents of the circles (26) and dividing by the total (52).

The Venn diagram is still more useful when events are *not* mutually exclusive. Suppose that event D is a picture card. There are 12 picture cards in a pack, and the probability of drawing a picture card $P(D) = 12/52$. If we now wish to find $P(A \text{ or } D)$, i.e. the probability that a card drawn at random is a heart or a picture, it would be incorrect to use:

$$P(A \text{ or } D) = P(A) + P(D)$$

since this would involve double counting – the heart pictures would be included in both $P(A)$ and $P(D)$. We need to deduct the heart pictures, i.e. we need to deduct $P(AD) = 3/52$. The additive law becomes:

$$P(A \text{ or } D) = P(A) + P(D) - P(AD)$$

Applying this result to our problem:

$$P(A \text{ or } D) = \frac{13}{52} + \frac{12}{52} - \frac{3}{52} = \frac{22}{52}$$

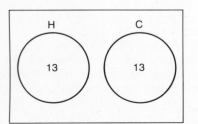

Fig. 9.2   Venn diagram

130

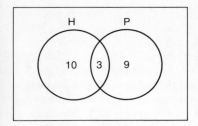

Fig. 9.3   Venn diagram

This result may also be obtained using a Venn diagram (see Fig. 9.3). The rectangle is the same as in Fig. 9.2 and contains 52 cards, but here the circles *overlap* in order to deal with the heart picture cards. The first step is to enter 3 in the overlap area; there are 9 other picture cards, and these should be entered into the remainder of the picture area. Similarly the 10 hearts *other than pictures* should be entered in the remainder of the heart area. The required probability that the card is a heart or a picture, is obtained by adding up these numbers 3 + 9 + 10 = 22 and then dividing by 52 to obtain the result found earlier. **Note:** When completing a Venn diagram it is essential to deal with the overlap area *first*.

Exercise 5

A large company normally recruits 20 accountancy trainees per annum. In 1978, 90 applications were received and of these:

> 63 had previous work experience;
> 36 had passed the foundation stage examination; and
> 27 had both work experience and had passed the foundation stage examination, and had been included in both the above counts.

(a) Prepare a Venn diagram to illustrate the above.
(b) What is the probability that an applicant taken at random:
    (i) had work experience, or had passed the foundation stage examination, or had both;
    (ii) had either work experience, or had passed the foundation examination, but not both?
(c) Given that the applicant must have work experience, determine the conditional probability that an applicant taken at random from this group will have passed the foundation stage examination.
(d) Given that the selected new recruits comprised 6 women and 14 men, state the probability, if 3 of the new recruits were chosen at random for a particular project, that they would all be male.

(ICMA May 1979)

(*Answers:* 0.8; 0.5; 0.4286; 0.3193)

Worked Example 6

In a country there are three Sunday newspapers; the *Echo*, *Advertiser* and *News*. A survey is conducted and it is found that 32% read the *Echo*; 38% read the *Advertiser*; 40% read the *News*; 7% read the *Echo* and the *Advertiser*; 9% read the *Advertiser* and the *News*; 8% read the *Echo* and the *News*; 3% read all three newspapers. Find:
(i) the percentage that read either the *Echo* or the *News* or both *Echo* and *News*;
(ii) the percentage that read at least one Sunday newspaper;
(iii) the percentage that read only the *Advertiser*.

This question is taken from an examination paper, but the question is ambiguous. It is not clear whether or not the 3% who read all three papers are included in the 7%, 9% or 8%; the solution assumes that they are included.

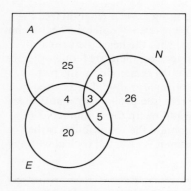

Fig. 9.4   Venn diagram

The Venn diagram is shown in Fig. 9.4, and we must begin with the areas of overlap. The first entry is the *triple* overlap of 3%. Next deal with the *double* overlaps. If 7% read the *Echo* and *Advertiser*, then $7-3=4\%$ read *Echo* and *Advertiser* but not the *News*. Similarly 9 $-3=6\%$ read the *Advertiser* and *News* but not the *Echo*, and $8-3$ $=5\%$ read the *Echo* and *News* but not the *Advertiser*. Finally, deal with areas with *no overlaps*. The percentage who read the *Echo* only is given by 32 less the overlap areas $(4+3+5)$ giving 20%. Similarly, those who read the *Advertiser* and *News* only are 25% and 26% respectively. We can now obtain the results:

(i)   Percentage is $20+26+6+3+4+5=64\%$.

(ii) Percentage is $20+25+26+6+3+4+5=89\%$.

(iii) Percentage is 25%.

(Worked example 5 was part of the text on Venn diagrams.)

Exercise 6

Your company decides to invite tenders for the supply of equipment for a new office. The company divides the equipment into three groups. Group A is office furniture (desks, chairs, etc.). Group B is information technology (word processors, computers, etc.). Group C is other equipment (filing cabinets, waste paper baskets, etc.).

In response, 26 firms tender for group A, 17 tender for group B, and 20 tender for group C. Of these firms, 5 tender for both group A and B but not C; 4 tender for both B and C but not A; 2 tender for A and C but not B; 3 firms tender for all three groups.

(a) How many firms tendered for one or more groups?

(b) How many firms tendered for one group only?

(c) One of the firms that tendered for all three groups considers that the probability that it will secure tender A is 0.2, tender B is 0.3 and tender C is 0.1. Find the probability that the firm

   (i)   will obtain all three contracts,

   (ii) will obtain only one contract.

(ICSA Dec. 1984)

(*Answers:* 46; 32; 0.006; 0.398)

C. SOLUTIONS TO EXERCISES

S1

$P(A \text{ or } B)=P(A)+P(B)$ $P(A \text{ or } B)=0.2+0.3=0.5$

S2

(i) $P(A)=\dfrac{12}{16}=0.75,\ P(3A)=0.75\times0.75\times0.75=0.4219$

(ii) $P(A)=\dfrac{12}{16},\ P(B/A)=\dfrac{11}{15},\ P(C/AB)=\dfrac{10}{14}$

$$P(ABC) = P(A) \times P(B/A) \times P(C/AB) = \frac{12}{16} \times \frac{11}{15} \times \frac{10}{14} = \frac{1,320}{3,360}$$
$$= 0.3929$$

P(Third item first good item) $= P$(First two defective) $\times P$(Third good)

$$P\text{(First two defective)} = \frac{4}{16} \times \frac{3}{15} = \frac{12}{240}, \ P\text{(Third good)} = \frac{12}{14},$$

thus result $= \dfrac{12}{240} \times \dfrac{12}{14} = \dfrac{144}{3,360} = 0.04286.$

S3

P(Good component) $= 1 - 0.01 = 0.99, \ P$(Toy good) $= 0.99^4 = 0.9606$
P(Toy defective) $= 1 - 0.9606 = 0.0394.$

S4

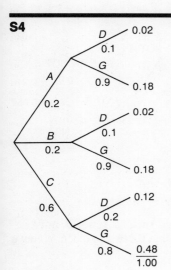

(i) As A and B are equally likely and C is three times as likely to be selected as B, then 60% of tubes are from C, and 20% from A or B. $P(A) = 0.2.$

(ii) See Fig. 9S.1. The probabilities have been entered on the tree. P(Defective) $= 0.02 + 0.02 + 0.12 = 0.16.$

(iii) P(Defective part from A) $= 0.02, \ P(A/D) = \dfrac{0.02}{0.16} = 0.125.$

Fig. 9S.1

S5

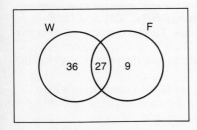

Fig. 9S.2

(a) See the Venn diagram in Fig. 9S.2.

(b) (i) From diagram the number is $36 + 27 + 9 = 72$;

Probability $= \dfrac{72}{90} = 0.8.$

(ii) Number is 45; Probability $= \dfrac{45}{90} = 0.5.$

(c) $P(F/D) = \dfrac{27}{63} = 0.4286$

(d) $P(3M) = \dfrac{14}{20} \times \dfrac{13}{19} \times \dfrac{12}{18} = \dfrac{2,184}{6,840} = 0.3193$

S6

For parts (a) and (b) see Fig. 9S.3. (a) is the sum of all the entries
$= 16 + 5 + 3 + 2 + 11 + 4 + 5 = 46$; (b) $16 + 11 + 5 = 32$

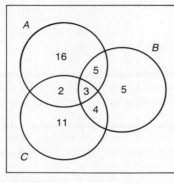

Fig. 9S.3

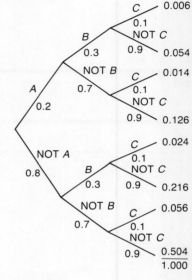

Fig. 9S.4

(c) (i)   From Fig. 9S.4, $P(ABC) = 0.006$.
    (ii) Here we need to find the branches such as A, not B, not C
       = 0.126 and add the other branches 0.216 and 0.056, giving a
       total of 0.398.

D. RECENT EXAMINATION QUESTIONS

Q1

(a) Three machines, A, B and C, produce 60%, 30% and 10%
respectively, of the total production. The percentages of defective
production of the machines are 2%, 4% and 6% respectively.
    If an item is selected at random, find the probability that the
item is defective.
(b) Using the information given in part (a) and assuming an item
selected at random is found to be defective, find the probability
that the item was produced on machine A.

(ICMA May 1976)

(*Answers:* 0.03; 0.4)

134

Q2
    (a) State clearly what is meant by two events being statistically independent.

    (b) In a certain factory which employs 10,000 men, 1% of all employees have a minor accident in a given year. Of these, 40% had safety instructions whereas 90% of all employees had no safety instructions. What is the probability of an employee being accident-free.

       (i) given that he had no safety instructions;

      (ii) given that he had safety instructions?

    (c) A company runs a special lottery. A box contains 100 tickets, an unknown number of which are winning tickets, that number having been selected from random number tables from integers 1 to 15 inclusive. A ticket is picked at random from the box. What is the probability that it is the winning ticket?

(ICMA May 1983)

(*Answers:* 0.9933; 0.96; 0.08)

Q3
    (a) Part 1 of the examinations of The Institute of Chartered Secretaries and Administrators consists of two modules. The subjects of Module 1 are: (i) Communication, (ii) General Principles of Law. The subjects of Module 2 are: (i) Principles of Economics, (ii) Statistics.

         A student attempts Part 1 of the examinations of the Institute of Chartered Secretaries and Administrators, i.e. he takes Modules 1 and 2. He considers that his chances of passing Communication is 0.7, General Principles of Law is 0.6, Principles of Economics is 0.8 and Statistics is 0.9. Assuming that the probability of his passing one subject is independent of the probability of his passing the other three subjects, find the probability:

       (i) that he passes Part 1,

      (ii) that he fails all four examinations,

     (iii) that he passes just one module, i.e. passes either Module 1 or Module 2.

    (b) Another student attempts Module 3. If the probability of passing this module is constant and equal to 0.6, find the probability that the students passes Module 3 at the third attempt.

(ICSA Dec. 1983)

(*Answers:* 0.3024; 0.0024; 0.5352; 0.096)

Q4
    (a) A company tenders for two contracts A and B. The probability that it will obtain contract A is 1/3 and the probability that it will obtain contract B is 1/4. Find the probability that the company:

       (i) will obtain both contracts

      (ii) will obtain only one contract.

    (b) A company has a large number of typists. A survey shows that 30 can use a word processor, 25 are audio-typists and 28 are shorthand writers. Of the typists who are shorthand writers, 3 are audio-typists and can use a word processor, 5 are audio-typists but cannot use a word processor, 9 can use a word processor but are not audio-typists. 6 of the audio-typists can use a word processor but are not shorthand writers.

(i)   Represent this information on a Venn diagram.
(ii) How many typists were involved in the survey?
(iii) How many typists have only one skill?

(ICSA Dec. 1982)

(*Answers:* 0.0833; 0.4167; 57; 34)

Q5

Employees have the choice of one of three schemes, *A*, *B* or *C*. They must vote for one but, if they have no preference, can vote for all three or, if against one scheme, they can vote for the two they prefer.

A sample poll of 200 voters revealed the following information:
15 would vote for *A* and *C* but not *B*;
65 would vote for *B* only;
51 would vote for *C* only;
15 would vote for both *A* and *B*;
117 would vote for either *A* or *B* or both *A* and *B* but not *C*;
128 would vote for either *B* or *C* or both *B* and *C* but not *A*.

How many would vote for: (a) all three schemes; (b) only one scheme; (c) *A* irrespective of *B* or *C*; (d) *A* only; (e) *A* and *B* but not *C*?

(ICMA Nov. 1977)

(*Answers:* 5; 158; 72; 42;10)

Q6

A manager in a department store has to decide how many luxury gift packs of cosmetics to buy for the forthcoming Christmas season. These gift packs have to be bought from a manufacturer in cases of 50. The profit per pack is £20.

The manager decides to use probability theory to aid her decision. The manager's estimate of sales of this gift pack this Christmas is as follows:

Sales (cases):	5	10	15	20	25
Probability:	0.2	0.2	0.3	0.2	0.1

Any unsatisfied demand does not affect the probability of future sales. Any unsold gift packs will be sold in the New Year Sale at a loss of £10 per pack.

(a)   Complete the following profit table (£'000), indicating any loss with a minus sign.

	Quantity of stock bought (cases), Q_j				
	5	10	15	20	25
Sales, S_i (cases) 5				−2.5	
10				5.0	
15				12.5	
20				20.0	
25				20.0	

(b) Assuming that the quantity of stock is bought in units of 5 cases, and ignoring storage costs, how many cases should the manager buy to maximize her expected profit?

Expected profit on $(Q_j) = \sum_i$ probability $(S_i) \times$ profit on (S_i).

(c) If the manager were completely uncertain about the probability of sales between 5 and 25 cases (inclusive), how many cases should she stock?

(ICMA May 1983)

(*Answers:* see Outline Answers for (a) and (c); (b) = 15)

E. OUTLINE ANSWERS TO EXAM QUESTIONS

A1

(a) See Fig. 9S.5. Probability is found by adding the branches with a D: $0.012 + 0.012 + 0.006 = 0.03$.

(b) $P(AD) = 0.012$

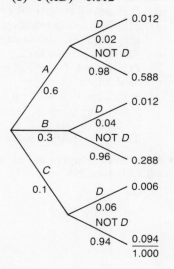

Fig. 9S.5

$$\text{Probability} = \frac{0.012}{0.03} = 0.4.$$

A2 (a) Two events are independent if the outcome of one event makes no difference to the outcome of the other event.

(b) The information can be placed in a table.
Number of accidents = 1% of 10,000 = 100.

	Safety inst.	No safety inst.	Total
Accident	40	60	100
No accident	960	8,940	9,900
Total	1,000	9,000	10,000

(i) $P(\text{No accident/no safety inst.}) = \dfrac{8,940}{9,000} = 0.9933$

(ii) $P(\text{No accident/safety inst.}) = \dfrac{960}{1,000} = 0.96$

(c) $P(\text{Winning ticket}) = \dfrac{\text{number of winning tickets}}{\text{number of tickets}}$

The number of winning tickets is between 1 and 15, the expected number of winning tickets is the mean of the numbers 1, 2, 3, ..., 15 which is 8,

Thus $P(\text{Winning ticket}) = \dfrac{8}{100} = 0.08.$

A3 (a) (i) $P(\text{Passes all exams}) = 0.7 \times 0.6 \times 0.8 \times 0.9 = 0.3024.$

(ii) $P(\text{Failing communication})$ is $1 - 0.7 = 0.3$; similarly for other subjects. $P(\text{Fails all exams}) = 0.3 \times 0.4 \times 0.2 \times 0.1 = 0.0024.$

(iii) Required probability is
$P(\text{Pass mod. 1}) \times P(\text{Fail mod. 2}) + P(\text{Pass mod. 2}) \times P(\text{Fail mod. 1})$
$P(\text{Pass mod. 1}) = 0.7 \times 0.6 = 0.42$; $P(\text{Fail mod. 1}) = 1 - 0.42 = 0.58$; $P(\text{Pass mod. 2}) = 0.8 \times 0.9 = 0.72$; $P(\text{Fail mod. 2}) = 1 - 0.72 = 0.28.$
Thus probability is $0.42 \times 0.28 + 0.72 \times 0.58 = 0.5352.$

(b) $P(\text{Pass 3rd attempt}) = P(\text{Fail twice}) \times P(\text{Pass at 3rd attempt}) = 0.4 \times 0.4 \times 0.6 = 0.096$

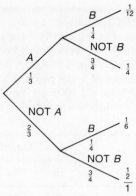

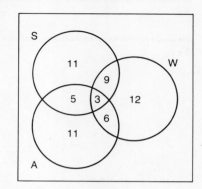

Fig. 9S.6

Fig. 9S.7

A4

(a) See Fig. 9S.6. From tree, (i) $1/12 = 0.0833$; (ii) $1/4 + 1/6 = 5/12 = 0.4167$.

(b) (i) Information has been entered on the Venn diagram; see Fig. 9S.7.

    (ii) The number of typists is found by adding all the entries $11 + 9 + 5 + 3 + 11 + 6 + 12 = 57$.

    (iii) Number with one skill only $= 11 + 11 + 12 = 34$.

A5

As far as practicable the information has been entered on a Venn diagram; see Fig. 9S.8. The letters w, x, y and z have been entered and need to be found. 15 would vote for A and B.

    Thus $x + y = 15$. . . (1)

117 would vote for either A or B or both A and B but not C.

    Thus $y + z + 65 = 117$

        $y + z = 52$ . . . (2)

128 would vote for either B or C or both B and C but not A.

    Thus $65 + w + 51 = 128$

        $w = 12$ . . . (3)

200 voted.

    Thus $65 + 15 + 51 + w + x + y + z = 200$

        $w + x + y + z = 69$ . . . (4)

Substitute from (3), $w = 12$, and from (1), $x + y = 15$, into (4):

    $12 + 15 + z = 69$ $z = 42$

Substitute $z = 42$ in (2): $y + 42 = 52$ $y = 10$

Substitute $y = 10$ in (1): $x + 10 = 15$ $x = 5$

(a) All three schemes $= x = 5$

(b) One scheme $= 65 + 51 + z = 158$

(c) A irrespective of B or $C = 15 + x + y + z = 72$.

(d) A only $= z = 42$.

(e) A and B but not $C = y = 10$.

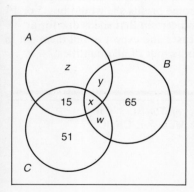

Fig. 9S.8

(a) If sales S_i equals or exceeds quantity bought Q_j there are no losses. For a quantity 5 the profits equal $5 \times 50 \times 20 = 5,000$, so 5 is entered at every point. For a quantity 10, if sales are 10 or more the profits equal $10 \times 50 \times 20 = 10,000$, so 10 should be entered. For sales of 5, the profit is $5 \times 50 \times 20(= 5,000)$ less the sales loss $5 \times 50 \times 10(= 2,500)$, giving a profit of 2,500, so 2.5 is entered. The other columns are found in the same way.

		Quantity bought (Q_j)				
		5	10	15	20	25
	5	5.0	2.5	0	−2.5	−5.0
	10	5.0	10.0	7.5	5.0	2.5
Sales (S_i)	15	5.0	10.0	15.0	12.5	10.0
	20	5.0	10.0	15.0	20.0	17.5
	25	5.0	10.0	15.0	20.0	25.0

(b) We take each column in turn and multiply the profit for the sale by the corresponding probability of the sale.

$Q = 5$: expected profit $= 5.0 \times (0.2 + 0.2 + 0.3 + 0.2 + 0.1)$
$= 5.0$

$Q = 10$: expected profit $= 2.5 \times 0.2 + 10.0 \times (0.2 + 0.3 + 0.2 + 0.1)$
$= 8.5$

$Q = 15$: expected profit $= 7.5 \times 0.2 + 15.0 \times (0.3 + 0.2 + 0.1)$
$= 10.5$

$Q = 20$: expected profit $= -2.5 \times 0.2 + 5.0 \times 0.2 + 12.5 \times 0.3$
$+ 20 \times (0.2 + 0.1) = 10.25$

$Q = 25$: expected profit $= -5.0 \times 0.2 + 2.5 \times 0.2 + 10.0 \times 0.3$
$+ 17.5 \times 0.2 + 25.0 \times 0.1 = 8.5$

The maximum expected profit occurs when $Q = 15$.

(c) The answer to this part depends on the criteria the manager follows. If she wants to ensure some profits she would buy 5 or 10 cases. If she were an optimist and certain that she is due for a lucky break, she would buy 25 cases and expect to sell them all. In fact an argument can be made in favour of all quantities!

A STEP FURTHER

Mulholland and Jones, *Fundamentals of Statistics*, Ch. 3.
Tennant-Smith, *Mathematics for the Manager*, Ch. 8.

Chapter 10

Normal, binomial and Poisson distributions

A. GETTING STARTED

You should first check whether all or any of these distributions are included in your syllabus. At the time of writing all three are included at the foundation level in the ICMA syllabus; none are included in the CACA (ACCA) syllabus and questions are not set on these distributions; none are specifically mentioned in the ICSA syllabus, but questions *are* set on the normal distribution and the examination paper includes an extract of the normal tables. If your syllabus includes significance testing and confidence intervals you should at least read the part of the chapter relating to the normal distribution.

B. ESSENTIAL PRINCIPLES

NORMAL DISTRIBUTION

The *normal distribution* is very important. The mean of large samples has a normal distribution, and this application of the normal distribution is covered in Chapter 11. Certain characteristics of data, such as heights of men (or women), intelligence quotients of children, demand for products etc., often follow a normal distribution.

The normal distribution with a mean μ and standard deviation σ is shown in Fig. 10.1. For data which is normally distributed, 68.3% of

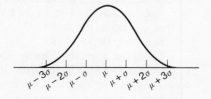

Fig. 10.1

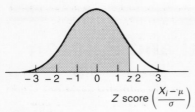

Fig. 10.2

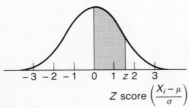

Fig. 10.3

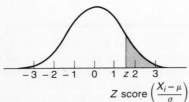

Fig. 10.4

observations are within one standard deviation of the mean; 95.4% of observations are within two standard deviations of the mean; and 99.7% of observations are within three standard deviations of the mean. To work out these percentages we need to find the areas under the normal curve. Special tables are available (and are usually provided in the examination room). These tables relate to a *standard normal distribution* with mean zero ($\mu = 0$) and standard deviation of unity ($\sigma = 1$). The Z 'score' or 'statistic' is defined as

$$Z = \frac{X_i - \mu}{\sigma},$$

where X_i is the observation, μ is the mean, and σ is the standard deviation. Clearly the Z score tells us the number of standard deviations any observation is from the mean of the distribution. Unfortunately there is no uniform method of presenting these tables. Different tables give different areas; Fig. 10.2, 10.3 and 10.4 illustrate these areas.

Tables giving the areas shaded in Fig. 10.2 and 10.3, i.e. to the *left* of the ordinate z, are the most common. The tables used in this book give the area in Fig. 10.2 (see p. 246). You must find out the type of tables used in your examination and practise using them.

The normal distribution is such that when z is on the extreme left of Fig. 10.2 the *area* to the left of Z is zero; when z is on the extreme right the *area* to the left of Z is unity, i.e. the whole area is to the left of Z. These properties of the normal distribution can be used to solve probability problems.

Worked Example 1

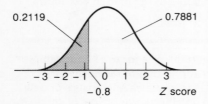

Fig. 10.5

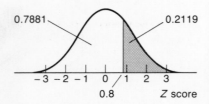

Fig. 10.6

Use the normal table on p. 246 to find the area (a) to the left, (b) to the right of z, if z is: (i) 1.5; (ii) 1.67; (iii) -0.8.

(i) From the table area to the left of 1.5 is 0.9332; area to the right is $1 - 0.9332 = 0.0668$.

(ii) The value for 1.67 is where the horizontal row from 1.6 meets the vertical column from 0.07, i.e. 0.9525; the area to the right is $1 - 0.9525 = 0.0475$.

(iii) To use the table on p. 246 for negative values, we have to note that the normal curve is symmetrical. The area to the left of -0.8 is the *same* as the area to the *right* of $+0.8$. In other words, our 'rule' is that the area to the left of a *negative* Z score is the same as *one minus* the area to the left of the equivalent *positive* Z score. See Figs. 10.5 and 10.6.

From the normal tables we find that the area to the *left* of 0.8 is 0.7881; thus the area to the *right* of 0.8 is $1 - 0.7881 = 0.2119$. It follows from our 'rule' that the area to the *left* of -0.8 is 0.2119, i.e. 1 $-$ the *area* value for the positive Z score. The area to the *right* of -0.8 is 0.7881.

Exercise 1

If z is normally distributed with mean 0 and standard deviation 1, find the probability that z is: (i) less than 1.8; (ii) greater than 0.85; (iii) is greater than -1.46.
(*Answers:* 0.9641; 0.1977; 0.9279)

Worked Example 2

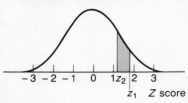

Fig. 10.7

If z is normally distributed with mean 0 and standard deviation 1, find the probability that z lies between: (i) 1.2 and 1.78; (ii) -0.35 and -2.45; (iii) -0.5 and 2.17.

It is helpful to draw a sketch; see Figs 10.7, 10.8 and 10.9.
   We need to find the probabilities (which means find the shaded areas since the total area under the curve is 1). In cases (i) and (ii) where the *areas* **do not** include the mean value 0, simply take the *difference* of the values found in the tables. In case (iii) where the *area* **does** include the mean value 0, the *area* value for the *negative Z* score (which is found by looking up the *positive Z* score in the tables) must be subtracted from 1 **before** taking differences.

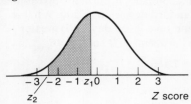

Fig. 10.8

(i) $z_1 = 1.78$, area $= 0.9625$; $z_2 = 1.2$, area $= 0.8849$. Required probability is $0.9625 - 0.8849 = 0.0776$.

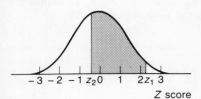

Fig. 10.9

(ii)   The area between -0.35 and -2.45 is the *same* as the area between $+0.35$ and $+2.45$. $z_1 = 0.35$, area $= 0.6368$; $z_2 = 2.45$, area $= 0.9929$. Required probability is $0.9929 - 0.6368 = 0.3561$.

(iii) $z_1 = 2.17$, area $= 0.9850$; $z_2 = -0.5$; for a positive value 0.5, area $= 0.6915$. As -0.5 is negative, the area $= 1 - 0.6915 = 0.3085$. Required probability is $0.9850 - 0.3085 = 0.6765$.

Exercise 2

If z is normally distributed with mean 0 and standard deviation 1, find the probability that z lies between: (i) 1.42 and 2.65; (ii) -1.45 and 2.31; (iii) -0.28 and -1.96.
(*Answers:* 0.0738; 0.9161; 0.3647)

Worked Example 3

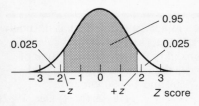

Fig. 10.10

If z is normally distributed with mean 0 and standard deviation 1 find the value of z if the area between $-z$ and $+z$ is 0.95.

The area required is shown in Fig. 10.10. If the shaded area is 0.95, the two tails have a combined area of 0.05 and so each tail has an area of 0.025. The area to the left of z is thus $0.95 + 0.025 = 0.975$. If we now look in the *body of the table* and find 0.975, we find that the corresponding value of z is 1.96.

| **Exercise 3** | Repeat Worked Example 3 but with an area of: (i) 0.9; (ii) 0.99. (*Answers:* 1.645; 2.575 – results to two decimal places are 1.64 and 2.58) |

NORMAL DISTRIBUTION WITH MEAN μ AND STANDARD DEVIATION σ

If we have such a distribution we need to *transform* the distribution so that we can use the tables. As we noted in deriving the Z score, we *standardize* by using:

$$z = \frac{x - \mu}{\sigma}$$

to transform to a standard normal distribution. The method is illustrated in Worked Example 4.

Worked Example 4

The daily demand for a product of a company is normally distributed with a mean of 2,000 and with a standard deviation of 200. What is the probability that the daily demand on a given day is: (i) greater than 2,450; (ii) less than 1,750; (iii) between 1,800 and 2,250?

(i) Here $\mu = 2,000$ and $\sigma = 200$; $x = 2,450$.

$$z = \frac{x - \mu}{\sigma} = \frac{2,450 - 2,000}{200} = \frac{450}{200} = 2.25.$$

From tables, area to the left of 2.25 is 0.9878; we require area to the right, so probability is $1 - 0.9878 = 0.0122$.

(ii) $x = 1,750$,

$$z = \frac{1,750 - 2,000}{200} = \frac{-250}{200} = -1.25.$$

From tables area to the left of $+1.25$ is 0.8944; the area to the left of $-1.25 = 1 - 0.8944 = 0.1056$.

(iii) $x = 1,800$, $x = 2,250$

$$z_1 = \frac{1,800 - 2,000}{200} = -1; \; z_2 = \frac{2,250 - 2,000}{200} = 1.25$$

From tables area to the left of $-1 = 1 - 0.8413 = 0.1587$; now area to the left of 1.25 is 0.8944; probability is $0.8944 - 0.1587 = 0.7357$.

Exercise 4

Daily electricity power consumption in a large office block is normally distributed with a mean of 10,000 kilowatts and a standard deviation of 2,000 kilowatts. What is the probability that the consumption of electricity on a given day is: (i) greater than 13,000 kilowatts, (ii) less than 8,000 kilowatts, (iii) between 7,500 and 14,000 kilowatts?

(ICSA (part) Dec. 1982)

(*Answers:* 0.0668; 0.1587; 0.8716)

Worked Example 5

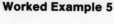

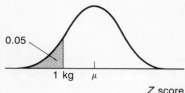

0.05

1 kg μ

Z score

Fig. 10.11

Sugar is packed in bags with a nominal weight of 1 kg, and the standard deviation of weights of bags of sugar is found to be 12 grams. The producer wishes to be sure that not less than 5% of the bags weigh less than 1 kg. What should be the mean weight of sugar bags assuming that weights are normally distributed?

If the mean is set at 1 kg, 50% would be underweight, so the mean has to be set above 1 kg. Figure 10.11 shows the position. We need to find μ so that the shaded area is 0.05. From the tables we need to find z corresponding to $1 - 0.05 = 0.95$. The Z score or value is 1.645, but we need a negative score -1.645.

$$x = 1, \sigma = 0.012, z = -1.645$$

$$z = \frac{x - \mu}{\sigma},$$

$$-1.645 = \frac{1 - \mu}{0.012}$$

$$\therefore \ \mu = 1 + 1.645 \times 0.012 = 1.0197 = 1.020 \text{ kg}$$

Exercise 5

Cartons of material are normally distributed with a mean weight of 520 grams and a standard deviation of 10 grams. If the supplier guarantees that the cartons will be a minimum weight of 500 grams, and needs to replace underweight cartons at a cost of £10 each, what is the likely cost of replacements following the supply of 10,000 cartons?

If the supplier wishes to reduce replacements to 1% or less, what mean weight should be fixed?
(*Answers:* £2,280; 523.3)

Factorial notation

5! is the mathematical symbol which means $5 \times 4 \times 3 \times 2 \times 1 = 120$. Some calculators have a key marked $x!$, if you enter 5 and press this key you obtain 120. Use your calculator to show that $6! = 720$.
Note: $1! = 1$ and $0! = 1$.

Permutations

This is the number of ways of *arranging items in order*. If, in a competition, you are given 10 desirable qualities for a car and are asked to place 4 of these in order of merit, in how many different ways can this be done? The first place can be chosen in 10 ways; having chosen the first, there are 9 left, so that the second place can be chosen in 9 ways. Similarly, the third place can be chosen in 8 ways, and the fourth place in 7 ways. The total number of ways is $10 \times 9 \times 8 \times 7 = 5,040$. This may be written using factorial notation

$$10 \times 9 \times 8 \times 7 = \frac{10 \times 9 \times 8 \times 7 \times 6 \times 5 \times 4 \times 3 \times 2 \times 1}{6 \times 5 \times 4 \times 3 \times 2 \times 1} = \frac{10!}{6!}$$

In *general*, if there are n items and r are to be placed *in order*, the number of ways is

145

$$_nP_r = \frac{n!}{(n-r)!}.$$

The example above would be written

$$_{10}P_4 \text{ i.e. } \frac{10!}{(10-4)!} = \frac{10!}{6!}$$

Combinations

Consider again the example used for permutations. Suppose we wish to select four desirable qualities of a car *irrespective of the order of merit*. The number of ways of doing this is equal to the number of permutations, 5,040, *divided by* the number of ways of placing the four selected items in order of merit. The latter can be done in 4! = 24 ways. The result, 210, is called the number of *combinations*. In terms of factorials this is equal to

$$\frac{10!}{6! \times 4!} = \frac{3,628,800}{720 \times 24} = 210$$

In *general*, the number of ways of choosing r items from n items, irrespective of order, is

$$_nC_r = \frac{n!}{(n-r)! \times r!}$$

It is useful to find $_nC_r$ for $r = 0, 1, 2, 3, \ldots$

$$r = 0: {}_nC_0 = \frac{n!}{n! \times 0!} = 1$$

$$r = 1: {}_nC_1 = \frac{n!}{(n-1)! \times 1!} = n$$

$$r = 2: {}_nC_2 = \frac{n!}{(n-2)! \times 2!} = \frac{n(n-1)}{2}$$

$$r = 3: {}_nC_3 = \frac{n!}{(n-3)! \times 3!} = \frac{n(n-1)(n-2)}{3 \times 2}$$

The values

$$1, n, \frac{n(n-1)}{2}, \frac{n(n-1)(n-2)}{3 \times 2}, \frac{n(n-1)(n-2)(n-3)}{4 \times 3 \times 2}, \ldots$$

are called *binomial coefficients*. We need these in the next section.

BINOMIAL DISTRIBUTION

The *binomial distribution* occurs when there are n *independent trials*, with the probability of success in each trial being constant. Let the probability of 'success' be equal to p, so that the probability of 'failure' in each trial is $1-p$, which is usually represented by q.

If we roll 4 unbiased dice and wish to find the probability of obtaining 0, 1, 2, 3, or 4 sixes, this requires the use of the binomial distribution; here $n = 4$ and $p = 1/6$. A manufacturer makes a product with a constant defective rate of 10%. If a sample of 5 items is taken from the production line, what is the probability that the sample contains 0, 1, 2, 3, 4, or 5 defectives? This is another example of the binomial distribution; here $n = 5$ and $p = 0.1$.

The probability of obtaining r successes in n trials is given by the formula:

$$P(r) = {}_nC_r\, p^r q^{n-r} = \frac{n!}{(n-r)! \times r!} p^r q^{n-r}$$

where $r = 0, 1, 2, 3, \ldots, n$

Using the results of the section on combinations, we find

$$P(0) = q^n, \qquad P(1) = npq^{n-1}, \qquad P(2) = \frac{n(n-1)}{2} p^2 q^{n-2}$$

$$P(3) = \frac{n(n-1)(n-2)}{3 \times 2} p^3 q^{n-3},$$

$$P(4) = \frac{n(n-1)(n-2)(n-3)}{4 \times 3 \times 2} p^4 q^{n-4}, \ldots$$

$$P(n-2) = \frac{n(n-1)}{2} p^{n-2} q^2, \qquad P(n-1) = np^{n-1}q, \qquad P(n) = p^n$$

If you are supplied with a formula list you need to check the form of presentation of the binomial distribution. If your calculator has a factorial key ! you can use the general formulae, but most students prefer to use the results for $P(0)$, $P(1)$, ... quoted above.

| **Worked Example 6** | If a manufacturer produces a product with a defective rate of 10% and if samples of 5 are taken from the production line, find the probability of 0, 1, 2, 3, 4, 5 defectives in the sample. |

Here $n = 5$, $p = 0.1$, $q = 1 - p = 0.9$. Substituting in the expressions above:

$$P(0) = q^5 = (0.9)^5 = 0.5905$$

$$P(1) = npq^4 = 5 \times 0.1 \times (0.9)^4 = 0.3281$$

$$P(2) = \frac{n(n-1)}{2} p^2 q^3 = \frac{5 \times 4}{2} \times (0.1)^2 \times (0.9)^3 = 0.0729$$

$$P(3) = \frac{n(n-1)(n-2)}{3 \times 2} p^3 q^2 = \frac{5 \times 4 \times 3}{3 \times 2} \times (0.1)^3 \times (0.9)^2 = 0.0081$$

$$P(4) = np^4 q = 5 \times (0.1)^4 \times 0.9 = 0.00045,$$

$$P(5) = p^5 = (0.1)^5 = 0.00001$$

| **Exercise 6** | 5% of the units produced in a manufacturing business are found to be defective. Find the probability that for a sample of 8 units 0, 1, 2, 3 units will be defective.
(*Answer:* 0.6634; 0.2793; 0.0514; 0.0054) |

If n and r are large, calculations such as those in Worked Example 6 above can become tedious. In such a situation the binomial distribution approximates to the *normal distribution* and we can use the normal tables to find the probabilities. It is possible to show that the mean and standard deviation of a binomial distribution are np and $\sqrt{npq}$ respectively. Worked Example 7 explains how to use the normal approximation.

Worked Example 7

20 coins are tossed and the number of heads r is noted. Find the probability that the number of heads is: (i) 15 or more; (ii) between 5 and 13 inclusive.

It is possible to follow the method of Worked Example 6; the results would be (i) 0.0207, (ii) 0.9364. This is the exact result.

(i)   A section of the histogram is shown in Fig. 10.12. If we wish to

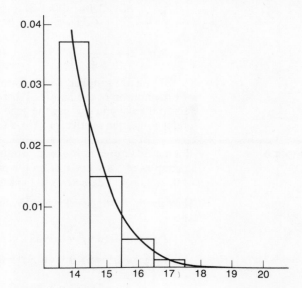

Fig. 10.12

use the normal approximation, we have to take into account that the normal curve is continuous whereas the binomial distribution takes integer (whole number) values only. To find the probability that the number of heads is 15 or more using the normal approximation, means that we have to find the probability of a value greater than 14.5. It is necessary to standardize $\mu = np$,

$\sigma = \sqrt{npq}$, $n = 20$, $p = 0.5$ $q = 1 - p = 0.5$; $\mu = 20 \times 0.5 = 10$,

$\sigma = \sqrt{20 \times 0.5 \times 0.5} = \sqrt{5} = 2.236$.

$$z = \frac{x - \mu}{\sigma} = \frac{14.5 - 10}{2.236} = \frac{4.5}{2.236} = 2.01$$

From the normal tables, the area to the left of 2.01 is 0.9778.
Required probability is $1 - 0.9778 = 0.0222$.

(ii) When using the normal approximation we need the probability of a value between 4.5 and 13.5.

$$z_1 = \frac{x - \mu}{\sigma} = \frac{4.5 - 10}{2.236} = -2.46 \qquad z_2 = \frac{13.5 - 10}{2.236} = 1.565$$

From the tables, the value corresponding to $+2.46$ is 0.9931, so area to left of -2.46 is 0.0069; the value corresponding to 1.565 is 0.9412 (taking the value mid-way between 0.9406 and 0.9418).
Required probability is $0.9412 - 0.0069 = 0.9343$.

The exact results are 0.0207 and 0.9364, so the normal approximations are reasonably close. *In general, the larger the value of n, the better the approximation.*

Exercise 7

A popular restaurant has places for 50 customers. For Sunday lunches there is great demand so it is necessary to book. The restaurant manager knows that 10% of customers who book do not arrive at the restaurant. He takes bookings for Sunday lunch for 55 customers. What is the probability that he will have more customers than places? (*Answer:* 0.3264)

POISSON DISTRIBUTION

The *Poisson distribution* may be regarded as a special case of the binomial distribution where n is *large* and p is *small*, but the mean np is constant. As p is small the chance of the event occurring is low. This distribution is used for accidents, strikes, etc.

If m is the mean, it is possible to show that the probability of r successes is given by the formula:

$$P(r) = \frac{e^{-m} m^r}{r!}$$

Where $r = 0, 1, 2, 3, \ldots$ and $e = 2.7183$
If we substitute $r = 0, 1, 2, 3, 4, \ldots$ in this formula we obtain:

$$P(0) = e^{-m}, \qquad P(1) = me^{-m}, \qquad P(2) = \frac{m^2 e^{-m}}{2},$$

$$P(3) = \frac{m^3 e^{-m}}{3 \times 2}, \qquad P(4) = \frac{m^4 e^{-m}}{4 \times 3 \times 2}, \qquad P(5) = \frac{m^5 e^{-m}}{5 \times 4 \times 3 \times 2}, \ldots$$

In examination questions you are given the mean m or you have to find m from the information given, which is usually data for n and p; m is then obtained from the relationship $m = np$.

In an examination situation you have to be able to work out e raised to a negative power. Tables supplied in the examination room may have a table for e^{-x}. Many calculators have a key marked e^x. The

149

easiest way to find (say) e^{-3} on your calculator is to enter 3, 'press $+/-$ key, press e key, and you should obtain 0.049787. If your calculator does not have an e key but has an x^y key, enter 2.7813, press x^y key, enter 3, press $+/-$ key, then press $=$ key; you should obtain 0.049786. If you own a basic calculator, find $2.7183 \times 2.7183 \times 2.7183 = 20.08594$, then divide 1 by $20.08594 = 0.049786$'.

Worked Example 8	Computational errors in an Accounts department have a mean of 4 per day. Calculate the probability that on a given day the number of errors is: (i) none; (ii) exactly 7; (iii) between 2 and 4 inclusive.

First step is to find $e^{-4} = 0.0183156$

(i) $P(0) = e^{-4} = 0.0183$

(ii) $P(7) = \dfrac{4^7 \times e^{-4}}{7!} = \dfrac{16,384 \times 0.0183156}{5,040} = 0.0595$

(iii) $P(2) = \dfrac{4^2 \times e^{-4}}{2!} = \dfrac{16 \times 0.0183156}{2} = 0.1465$

$\quad\quad P(3) = \dfrac{4^3 \times e^{-4}}{3!} = \dfrac{64 \times 0.0183156}{6} = 0.1954$

$\quad\quad P(4) = \dfrac{4^4 \times e^{-4}}{4!} = \dfrac{256 \times 0.0183156}{24} = 0.1954$

Probability $= 0.1465 + 0.1954 + 0.1954 = 0.5373$

Exercise 8	Bus-Hire Ltd has two coaches which it hires out for local use by the day. The number of demands for a coach on each day is distributed as a Poisson distribution, with a mean of two demands. (i) On what proportion of days is neither coach used? (ii) On what proportion of days is at least one demand refused? (iii) If each coach is used an equal amount, on what proportion of days is **one** particular coach not in use? (ICMA (part) Nov. 1981) (*Answers:* 0.1353; 0.3233; 0.2707)

C. SOLUTIONS TO EXERCISES

S1	(i) Tables give 0.9641; (ii) tables give 0.8023, as *greater than* 1.8; subtract from $1 = 0.1977$; (iii) area same as *less than* $+1.46 = 0.9279$.

S2	(i) From tables 2.65 gives 0.9960, 1.42 gives 0.9222; probability is $0.9960 - 0.9222 = 0.0738$; (ii) area to left of 2.31 is 0.9896, area to left of -1.45 is $1 -$ area to left of $+1.45 = 0.0735$; probability is $0.9896 - 0.0735 = 0.9161$; (iii) find area between $+0.28$ and $+1.96$; $0.9750 - 0.6103 = 0.3647$.

S3

(i) If central area is 0.9, each tail is 0.05, area to left of z is 0.95; 1.64 gives 0.9495, 1.65 gives 0.9505; 0.95 is midway so z is 1.645.

(ii) For 0.99 each tail is 0.005, area to left of z is 0.995; 2.57 gives 0.9949, 2.58 gives 0.9951, thus z is 2.575.

S4

(i) $z = \dfrac{13{,}000 - 10{,}000}{2{,}000} = 1.5.$

From tables area to left is 0.9332, probability is $1 - 0.9332 = 0.0668$.

(ii) $z = \dfrac{8{,}000 - 10{,}000}{2{,}000} = -1.$

Area to left of $+1$ is 0.8413, probability is $1 - 0.8413 = 0.1587$.

(iii) $z_1 = \dfrac{7{,}500 - 10{,}000}{2{,}000} = -1.25, \quad z_2 = \dfrac{14{,}000 - 10{,}000}{2{,}000} = 2$

Area to left of $-1.25 = 1 - 0.8944 = 0.1056$, area to left of 2 is 0.9773; probability is $0.9773 - 0.1056 = 0.8716$.

S5

$$z = \frac{500 - 520}{10} = -2;$$

area to left of -2 is $1 - 0.9772 = 0.0228$. A proportion 0.0228 underweight out of 10,000 is 228, replacement cost $228 \times 10 = £2{,}280$. If replacements to be 1% or less, mean has to be larger. Value of z corresponding to 0.99 is 2.33. Now $-2.33 = \dfrac{500 - \mu}{10}$, so $\mu = 500 + 2.33 \times 10 = 523.3$.

S6

$n = 8$, $p = 0.05$, $q = 0.95$, $P(0) = (0.95)^8 = 0.6634$

$P(1) = 8 \times (0.95)^7 \times (0.05) = 0.2793, \quad P(2) = \dfrac{8 \times 7}{2} \times (0.95)^6 \times (0.05)^2$

$= 0.0514, \quad P(3) = \dfrac{8 \times 7 \times 6}{3 \times 2} \times (0.95)^5 \times (0.05)^3 = 0.0054.$

S7

Here $n = 55$, $p = 0.9$, $\mu = np = 49.5$, $\sigma = \sqrt{npq} = 2.225$. We require probability that 51 or more customers appeared.

$$z = \frac{50.5 - 49.5}{2.225} = 0.45.$$

Area to left of 0.45 is 0.6736, probability is $1 - 0.6736 = 0.3264$.

S8

$m = 2, e^{-2} = 0.1353352$.

(i) $P(0) = e^{-2} = 0.1353$.

(ii) $P(r > 2) = 1 - P(0) - P(1) - P(2)$. $P(1) = me^{-2} = 2 \times 0.1353352$
$= 0.2707$,

$$P(2) = \frac{2^2 \times 0.1353352}{2} = 0.2707.$$

Probability is $1 - 0.1353 - 0.2707 - 0.2707 = 0.3233$.

(iii) A coach is not in use when demand is zero and half the time when the demand is 1. Proportion is $P(0) + 0.5 \times P(1) = 0.1353 + 0.5 \times 0.2707 = 0.2707$.

D. RECENT EXAMINATION QUESTIONS

Q1

The length of life of batteries for electronic calculators is normally distributed with a mean of 1,000 hours with a standard deviation of 200 hours. Calculate the probability that the length of life of a battery is: (i) less than 700 hours; (ii) more than 1,400 hours; (iii) betwen 650 and 1,250 hours.

(ICSA (part) June 1985)

(*Answers:* 0.0668; 0.0228; 0.8543)

Q2

(a) An airline deliberately overbooks its local Mini flights because it knows from experience that not all passengers who book for a given flight actually arrive for that flight.

It is assumed that the probability of any booked passenger arriving for a given flight is 0.8; this is independent of the probability of any other passenger arriving.

The airline takes ten bookings for an 8-seater aircraft. Use the binomial distribution to find the probability for a given flight:

(i) that the aircraft takes off full;

(ii) that the aircraft takes off with at least two empty seats.

(b) The independent probability of a passenger arriving for a booked flight on a Maxi service is 0.8. The airline books 225 passengers and there are 195 seats available on a Maxi.

Use a normal distribution approximation to find the probability that for a given flight more booked passengers arrive than there are seats available.

(ICMA Nov. 1983)

(*Answers:* 0.6778; 0.1209; 0.0049)

Q3

Experience has shown that, on average, 2% of an airline's flights suffer a minor equipment failure in an aircraft. Use the Poisson distribution

152

to estimate the probability that the number of minor equipment failures in the next 50 flights will be (i) zero; (ii) at least two.

(ICMA (part) May 1982)

(*Answers:* 0.3679; 0.2642)

Q4

A machine in a factory has been set up to manufacture components normally distributed around a mean length of 60 mm. All components within 0.09 mm of the mean length are acceptable. At present 3.58% of components being produced have to be rejected because they are either too long or too short. Each component has a variable cost of £50.

(a) What is the standard deviation for the length of components currently being produced?

(b) It is possible to adapt the machine in order to reduce the variability of the lengths produced. The new standard deviation would be 0.0333 mm. What percentage of components would be rejected if the machine were to be adapted?

(c) Adapting the machine would increase fixed costs by £20,000 per annum. What is the minimum annual production of components which would make the adaption worthwhile?

(ABE Dec. 1983)

(*Answers:* 0.0429; 0.7; 13,889)

Q5

In a certain large factory the mean number of stoppages per week is 1.5. What is the probability that:

(i)   in a given week there will be no stoppages;

(ii) in a given week there will be three or more stoppages;

(iii) in a given two-week period there will be at most one stoppage?

(ICMA (part) May 1983)

(*Answers:* 0.2231; 0.1912; 0.1992)

Q6

(a) A company produces batteries whose lifetimes are normally distributed with a mean of 100 hours. It is known that 90% of batteries last at least 40 hours.

(i) Estimate the standard deviation lifetime.

(ii) What percentage of batteries will not last 70 hours?

(b) A company mass-produces electronic calculators. From past experience it knows that 90% of the calculators will be in working order and 10% will be faulty if the production process is working satisfactorily. An inspector randomly selects 5 calculators from the production line every hour and carries out a rigorous check.

(i) What is the probability that a random sample of 5 will contain at least 3 defective calculators?

(ii) A sample of 5 calculators is found to contain 3 defectives; do you consider the production process to be working satisfactorily?

(ICMA Nov. 1982)

(*Answers:* 46.88; 26.1; 0.00856)

Q7

    (a) Your company requires a special type of inelastic rope which is available from only two suppliers. Supplier A's ropes have a mean breaking strength of 1,000 kg with a standard deviation of 100 kg. Supplier B's ropes have a mean breaking strength of 900 kg with a standard deviation of 50 kg. The distribution of the breaking strengths of each type of rope is normal. Your company requires that the breaking strength of a rope be not less than 750 kg. All other things being equal, which rope should you buy, and why?

    (b) 1% of calculators produced by a company is known to be defective. If a random sample of 50 calculators is selected for inspection, calculate the probability of getting no defectives by using: (i) the binomial distribution; (ii) the Poisson distribution.

                                              (ICMA May 1984)

(*Answers: B*; 0.6050; 0.6065)

Q8

A manufacturer wishes to produce a new line of men's leisure shirts in five sizes, ranging from 'extra small' to 'extra large', i.e. XS, S, M, L and XL. From previous experience it is expected that demand will follow a normal distribution with a mean chest size of 90 cm and a standard deviation of 10 cm. For production reasons the manufacturer wishes to produce equal numbers of each size, subject to the restriction of catering for only the main 95% of demand.

    (i)   Find the upper and lower limits (in centimetres) of each of the five sizes which will meet the manufacturer's requirements.

    (ii) Explain briefly any advantages and disadvantages of trying to cater for the main 99% of demand, instead of the main 95%.

                                             (ICMA Nov. 1984)

(*Answers:* 70.4 to 82.1; 82.1 to 87.6; 87.6 to 92.4; 92.4 to 97.9; 97.9 to 109.6)

E. OUTLINE ANSWERS TO EXAM QUESTIONS

A1

$\mu = 1,000$, $\sigma = 200$.

    (i) $z = \dfrac{700 - 1,000}{200} = -1.5$,

           probability is $1 - 0.9332 = 0.0668$.

    (ii) $z = \dfrac{1,400 - 1,000}{200} = 2$,

           probability is $1 - 0.9772 = 0.0228$.

    (iii) $z_1 = \dfrac{650 - 1,000}{200} = -1.75$, $z_2 = \dfrac{1,250 - 1,000}{200} = 1.25$.

           Probability is $0.8944 - (1 - 0.9599) = 0.8543$.

A2 (a) $n = 10$, $p = 0.8$, $q = 0.2$.
  (i) We need $P(8) + P(9) + P(10)$:

$$P(8) = \frac{n(n-1)}{2}p^8q^2 = \frac{10 \times 9}{2} \times (0.8)^8 \times (0.2)^2 = 0.3020;$$

$$P(9) = np^9q = 10 \times (0.8)^9 \times 0.2 = 0.2684;$$
$$P(10) = p^{10} = (0.8)^{10} = 0.1074.$$
Probability $= 0.3020 + 0.2684 + 0.1074 = 0.6778$.
  (ii) We require $P(r < 6)$ which is $1 - P(7) - P(8) - P(9) - P(10)$

$$P(7) = \frac{n(n-1)(n-2)}{3 \times 2}p^7q^3 = \frac{10 \times 9 \times 8}{3 \times 2}(0.8)^7 \times (0.2)^3 = 0.2013.$$

Probability is $1 - 0.2013 - 0.6778 = 0.1209$.

(b) $n = 225$, $p = 0.8$, $q = 0.2$, $np = 180$, $\sigma = \sqrt{npq} = 6$. We need probability $x > 195.5$.

$$z = \frac{195.5 - 180}{6} = 2.58;$$

probability is $1 - 0.9951 = 0.0049$.

A3 $n = 50$, $p = 0.02$, $m = np = 1$. (i) $P(0) = e^{-m} = 0.3679$; (ii) $P(r \geqslant 2) = 1 - P(0) - P(1)$. $P(1) = me^{-m} = 1 \times 0.3679 = 0.3679$; probability is $1 - 2 \times 0.3679 = 0.2642$.

A4 (a) If 3.58% rejected then 1.79% are too long. Proportion less than 60.09 is $1 - 0.0179 = 0.9821$, the corresponding value of z is 2.1.

$$z = \frac{x - \mu}{\sigma}, \; 2.1 = \frac{60.09 - 60}{\sigma}, \; \sigma = \frac{0.09}{2.1} = 0.0429.$$

(b) $z_1 = \frac{60.09 - 60}{0.0333} = 2.70$, $z_2 = \frac{59.91 - 60}{0.0333} = -2.70$.

From tables area is $0.9965 - (1 - 0.9965) = 0.993$, percentage rejected $= 100 \times (1 - 0.993) = 0.7$.

(c) Rejection percentage reduced from 3.58 to $0.7 = 2.88$. Let A be annual production, then

$$\frac{2.88}{100} \times A \times 50 = 1.44 \times A$$

is the reduction in cost of rejects. For break-even this must be equal to 20,000 thus break-even point is 13,889.

A5 $m = 1.5$, $e^{-1.5} = 0.2231301$.
  (i) $P(0) = e^{-m} = 0.2231$.

(ii) $P(r \geqslant 3) = 1 - P(0) - P(1) - P(2)$, $P(1) = me^m = 1.5 \times 0.2231301$
$= 0.3347$,

$$P(2) = \frac{m^2 e^{-m}}{2} = \frac{(1.5)^2}{2} \times 0.2231301 = 0.2510;$$

$P(r \geqslant 3) = 1 - 0.2231 - 0.3347 - 0.2510 = 0.1912$.

(iii) Possibilities are no stoppages in both weeks $= 0.2231^2 = 0.0498$;
one stoppage in week 1 but none in week 2 $= 0.2231 \times 0.3347$
$= 0.0747$; no stoppage in week 1 but one stoppage in week 2
$= 0.3347 \times 0.2231 = 0.0747$. Probability $= 0.0498 + 2 \times 0.0747$
$= 0.1992$.

A6

(a) (i) 10% last less than 40 hours. We need value of z
corresponding to -0.1. Look up 0.9 in tables; value of z is
1.28, but take negative value -1.28, thus

$$-1.28 = \frac{40 - 100}{\sigma}, \quad \sigma = \frac{60}{1.28} = 46.88.$$

(ii) $z = \dfrac{70 - 100}{46.88} = -0.6399$.

Tables give 0.7389 for $+0.64$; result is $1 - 0.7389 = 0.2611$
$= 26.1\%$.

(b) (i) $n = 5$, $p = 0.1$, $q = 0.9$. $P(r \geqslant 3) = P(3) + P(4) + P(5)$.

$$P(3) = \frac{n(n-1)}{2} p^3 q^2 = \frac{5 \times 4}{2} \times (0.1)^3 \times (0.9)^2 = 0.0081;$$

$P(4) = np^4 q = 5 \times (0.1)^4 \times 0.9 = 0.00045$ $P(5) = p^5 = (0.1)^5$
$= 0.00001$. $P(r \geqslant 3) = 0.0081 + 0.00045 + 0.00001 = 0.00856$.

(ii) Probability of 3 defectives is 0.0081; if the production process
is working normally we would expect 3 defectives on 81
occasions out of 10,000 – it is more likely that defective rate
has increased above 10%.

A7

(a) Probability that ropes will break:

for A, $z = \dfrac{750 - 1,000}{100} = -2.5$.

Tables give $1 - 0.9938 = 0.0062$.

For B, $z = \dfrac{750 - 900}{50} = -3$.

Tables give $1 - 0.9987 = 0.0013$. Rope B has a higher percentage
that will not break over 750 kg and is to be preferred.

(b) $n = 50$, $p = 0.01$, $q = 0.99$. (i) Binomial, $P(0) = q^{50} = 0.6050$;
(ii) Poisson $m = np = 50 \times 0.01 = 0.5$, $P(0) = e^{-0.5} = 0.6065$.

(i) As 95% to be covered, we need to use 1.96, as obtained in Worked Example 3.

$$1.96 = \frac{x - \mu}{\sigma}, \ 1.96 = \frac{x - 90}{10}, \ x = 109.6,$$

which is largest chest size; by symmetry smallest chest size is 70.4. We are required to have five equal sizes. As we are catering for 95% each size covers 19%. Let us deal with largest size; the boundary is at $19 + 2.5 = 21.5\%$. We use the normal tables to find z corresponding to $1 - 0.215 = 0.785$ which is

$$z = 0.79; \ z = \frac{x - \mu}{\sigma}, \ 0.79 = \frac{x - 90}{10}, \ x = 97.9,$$

thus largest size is 97.9 to 109.6. By symmetry smallest size is 70.4 to 82.1. To find next largest size, the boundary is at $21.5 + 19 = 40.5\%$. From the normal tables z corresponding to $1 - 0.405 = 0.595$ is 0.24;

$$z = \frac{x - \mu}{\sigma}, \ 0.24 = \frac{x - 90}{10}, \ x = 92.4.$$

Thus size L is 92.4 to 97.9; by symmetry size S is 82.1 to 87.6. The remaining size M is 87.6 to 92.4.

(ii) If 99% the range of chest measurements is 64.2 to 115.8, so size ranges will become larger, but more of the demand will be met.

A STEP FURTHER

Mulholland and Jones, *Fundamentals of Statistics*, Chs 4 and 8. Tennant-Smith, *Mathematics for the Manager*, Ch. 8.

Chapter 11

Confidence intervals and significance testing

A. GETTING STARTED

You should first check whether these topics are included in your syllabus. When quantitative courses at foundation level were solely on statistics, these topics were almost always included. Now that the quantitative courses at foundation level often include mathematics and/or computing and data processing, the topics of this chapter are often omitted. At the time of writing the ICMA and CACA (ACCA) exclude these topics, although they are part of most diploma and undergraduate courses with options in quantitative methods.

B. ESSENTIAL PRINCIPLES

STATISTICAL INFERENCE

There are many practical situations where we have to use data from a sample to obtain information. The Department of Employment needs to find what the average household spends on goods and services in order to work out the weights of the index of retail prices. It would be impracticable to question *every* household in the United Kingdom, so the basic data is obtained from a survey of about 7,000 households (the Family Expenditure Survey). We use data from (say) how much the average family in the survey spends on food to estimate the figure for the whole country.

A general election in the United Kingdom is held every four or five years, so, at the time of the election, we know the proportions who voted for each of the main parties. At other times the only estimate of support for the main parties can be obtained from an opinion poll. The opinion poll is usually based on a random sample of about 1,000 voters and this is used to estimate the voting intentions of the electorate of more than 40 million.

An educational institution wishes to examine the effectiveness of computer-assisted learning in the teaching of economics. It could take a random sample of economics students, and split the sample into two groups: one group follows a computer-assisted course in economics and the other group follows the usual non-computer assisted course. At the end of the respective courses both groups take an examination in economics. Do the results of the examination support the theory that computer-assisted learning is superior to traditional methods? In this situation we are testing a hypothesis.

NOTATION

We use *sample* data to estimate *population* data. For example, we use the mean expenditure on food from the *sample* of households to estimate the mean expenditure on food of the population, i.e. *all* households in the United Kingdom. We need a system of notation to *differentiate* between a sample value and a population value. The convention is to use a Roman letter to represent the sample value and a Greek letter to represent the population value. Thus $\bar{x}$ is the *sample mean*, which is an estimate of μ the *population mean*; s is the *sample standard deviation* which is an estimate of σ the *population standard deviation*; p is the *sample proportion* which is an estimate of π the *population proportion*.

SAMPLING DISTRIBUTION

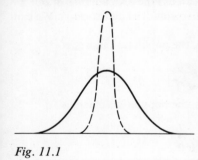

Consider the experiment to examine the effectiveness of computer-assisted learning. When we take a sample of students, the sample may or may not be representative of the whole student body. If a particular sample were to contain *an above average* proportion of very intelligent students then a high mean score in an examination may be due to that sample having intelligent students. If we take a *number* of samples of students, then we would expect the sample mean $\bar{x}$ to vary from sample to sample, simply because some samples may contain a high proportion of intelligent students whereas other samples may contain a low proportion of intelligent students. We say that the sample mean $\bar{x}$ has a *sampling distribution*. It is possible to show that in general the sample mean $\bar{x}$ has a normal distribution with mean μ and standard deviation $\sigma/\sqrt{n}$ ($\sigma/\sqrt{n}$ is called the *standard error*, instead of standard deviation, when the distribution refers to *samples*).

It does not matter whether the population is normal or skew. The *central limit theorem* (the central limit theorem is outside the syllabus of many courses at the introductory level) shows that, in general, the sample mean $\bar{x}$ has a normal distribution provided n is *large*, i.e. greater than 30. Figure 11.1 shows a *normal* distribution for the *population*; Fig. 11.2 shows a *skew* distribution for the *population*. In both figures the dotted curve shows the *sampling distribution* of $\bar{x}$.

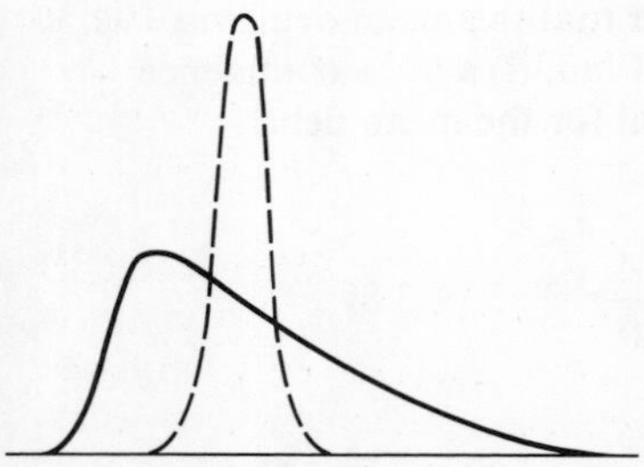

The fact that the distribution of the *sample mean* is normal allows us to use some of the results of Chapter 10.

Fig. 11.1

Fig. 11.2

CONFIDENCE INTERVAL FOR A MEAN

Suppose we take a sample of students and give them an intelligence test and find the sample mean $\bar{x}$. Then, if the sample contains a high proportion of intelligent students, the sample mean $\bar{x}$ will be greater than the population mean μ. If, however, the sample contains a low proportion of intelligent students, then the simple mean $\bar{x}$ will be less than the population mean μ. If the sample is representative of the population, then the sample mean $\bar{x}$ is likely to be close to the population mean μ.

When we take a sample we have no idea whether the *particular* sample is representative of the population or not. To allow for the fact that the sample may or may not be representative, it is the usual procedure to give *an interval* which specifies the limits within which it is likely that the population mean will lie. Such an interval is called a *confidence interval*.

In Worked Example 3 in Chapter 10 we found that 0.95 of the population (i.e. 95% of the entire observations of a normal distribution with mean 0 and standard deviation 1) lies between the z scores of -1.96 and $+1.96$. Put another way, 0.95 of the population lies within 1.96 standard deviations of the mean. The *sample mean $\bar{x}$* also has a normal distribution, according to the central limit theorem, provided the sample size $n > 30$. The mean of the *sampling distribution* is equal to the *population* mean, μ, but the standard error (*not* deviation, remember) is different. In fact, the standard error is $\sigma/\sqrt{n}$ and *not* σ. We often use the *sample standard deviation*, s, as an *estimate* of σ, in which case the standard error is $s/\sqrt{n}$. Because the sampling distribution is normal for $n > 30$, we can use the probabilities of our z score and normal tables (mean zero and standard deviation 1). We can therefore state the following.

A 95% confidence interval for μ, the population mean, is

$$\bar{x} - 1.96 \frac{s}{\sqrt{n}} \quad \text{to} \quad \bar{x} + 1.96 \frac{s}{\sqrt{n}}$$

A 99% confidence interval for μ, the population mean, is

$$\bar{x} - 2.58 \frac{s}{\sqrt{n}} \quad \text{to} \quad \bar{x} + 2.58 \frac{s}{\sqrt{n}}$$

Worked Example 1

A random sample of 50 debts showed that the mean debt was £92.50 with a standard deviation of £14.05. Find: (i) a 95% confidence interval; (ii) a 99% confidence interval for the mean debt.

$\bar{x} = 92.50$, $s = 14.05$, $n = 50$

(i) $\quad \bar{x} + 1.96 \dfrac{s}{\sqrt{n}} = 92.50 + 1.96 \times \dfrac{14.05}{\sqrt{50}} = 92.50 + 3.89 = 96.39$

$\quad\quad \bar{x} - 1.96 \dfrac{s}{\sqrt{n}} = 92.50 - 1.96 \times \dfrac{14.05}{\sqrt{50}} = 92.50 - 3.89 = 88.61$

95% confidence interval is £88.61 to £96.89.

(ii) $\bar{x} + 2.58 \dfrac{s}{\sqrt{n}} = 92.50 + 2.58 \times \dfrac{14.05}{\sqrt{50}} = 92.50 + 5.13 = 97.63$

$\bar{x} - 2.58 \dfrac{s}{\sqrt{n}} = 92.50 - 2.58 \times \dfrac{14.05}{\sqrt{50}} = 92.50 - 5.13 = 87.37$

99% confidence interval is £87.37 to £97.63.

Exercise 1

A sample of 200 households buying a house with a mortgage showed that the average monthly payment was £175.72 with a standard deviation of £42.60. Find: (i) 95%; (ii) 99% confidence intervals for mean payment.
(*Answers:* £169.82 to £181.62; £167.95 to £183.49)

CONFIDENCE INTERVAL, FOR A PROPORTION

Suppose that a sample survey shows that the proportion of persons buying a house by mortgage who are under 30 is 35%; can we find a 95% confidence interval for the whole population? We may show that if the sample is large ($n > 30$) then the *sample proportion p* also has a normal distribution with mean π and standard deviation

$$\sqrt{\dfrac{p(1-p)}{n}}.$$

A 95% confidence interval for π is:

$$p - 1.96 \sqrt{\dfrac{p(1-p)}{n}} \quad \text{to} \quad p + 1.96 \sqrt{\dfrac{p(1-p)}{n}}$$

A 99% confidence interval for π is:

$$p - 2.58 \sqrt{\dfrac{p(1-p)}{n}} \quad \text{to} \quad p + 2.58 \sqrt{\dfrac{p(1-p)}{n}}$$

Worked Example 2

A large company takes a sample of 200 employees and finds that 30% travel more than 10 miles to work. Find: (i) 95%; (ii) 99% confidence intervals for the percentage of employees of the company who travel more than 10 miles to work.

$$p = \dfrac{30}{100} = 0.3, \; n = 200$$

(i) $\quad p - 1.96 \sqrt{\dfrac{p(1-p)}{n}} \quad \text{to} \quad p + 1.96 \sqrt{\dfrac{p(1-p)}{n}}$

$$= 0.3 - 1.96 \sqrt{\dfrac{0.3(1-0.3)}{200}} \quad \text{to} \quad 0.3 + 1.96 \sqrt{\dfrac{0.3(1-0.3)}{200}}$$

161

$$= 0.3 - 0.0635 \quad \text{to} \quad 0.3 + 0.0635 = 0.2365 \text{ to } 0.3635$$

$$= 23.65\% \text{ to } 36.35\%.$$

(ii) $\quad p - 2.58 \sqrt{\dfrac{p(1-p)}{n}} \quad \text{to} \quad p + 2.58 \sqrt{\dfrac{p(1-p)}{n}}$

$$= 0.3 - 2.58 \sqrt{\dfrac{0.3(1-0.3)}{200}} \quad \text{to} \quad 0.3 + 2.58 \sqrt{\dfrac{0.3(1-0.3)}{200}}$$

$$= 0.3 - 0.0836 \quad \text{to} \quad 0.3 + 0.0836 = 0.2164 \text{ to } 0.3836$$

$$= 21.64\% \text{ to } 38.36\%.$$

Exercise 2

A large company takes a random sample of 1,000 accounts and finds that 200 have been in arrears for more than 3 months. Find a 95% confidence interval for the percentage of accounts in arrears. If the company has a total of 50,000 accounts, find a 95% confidence interval for the number of accounts in arrears.
(*Answer:* 17.52 to 22.48; 8,760 to 11,240)

SIGNIFICANCE TEST FOR A SINGLE MEAN

Earlier we considered whether computer-assisted learning was effective in the teaching of economics. Suppose that the mean score in a multiple-choice test in economics taught conventionally gave a score of μ_0. We now take a random sample of n students and teach them economics using a specially designed computer program. These students are then given the same multiple-choice test, and we note the mean score $\bar{x}$ and the standard deviation s.

The sample mean $\bar{x}$ is the estimate of the population mean score μ of *those taught by the computer program*. We wish to examine the *null* (or initial) hypothesis that $\mu = \mu_0$, i.e. that there is no difference in the results from the two teaching methods. The *alternative* hypothesis is that $\mu \neq \mu_0$ (the sign $\neq$ means not equal to). This means that there *is* a difference in the teaching methods. We have to compare $\bar{x}$ with μ_0. Now $\bar{x}$ can be influenced by factors other than the effect of computer-assisted learning; for example, the sample of students selected for the computer-assisted course could by chance have had a high proportion of intelligent students. In this situation we would expect such a sample to have a higher mean score. What we require is a procedure for examining whether the difference between the sample mean $\bar{x}$ and μ_0 is likely to be due to:

either

(i) The fact that the sample was unrepresentative, i.e. contained a high (or low) proportion of intelligent students;

or

(ii) That computer-assisted learning really did make a difference.

We must employ a *significance test* if we wish to find out whether or not any observed difference between $\bar{x}$ and μ_0 is *significant*. If the result *is* deemed significant, we conclude that (ii) is most likely, i.e. that computer-assisted learning has made a difference. If the result is deemed *not* significant, we conclude that any difference could be explained by the sample being unrepresentative.

We shall use the fact that the *difference* between $\bar{x}$ and μ_0 divided by the standard error $s/\sqrt{n}$, is normally distributed with mean 0 and standard deviation 1. We can then use the normal tables:

i.e. $z = \dfrac{\bar{x} - \mu_0}{s/\sqrt{n}}$ is normally distributed.

ONE-TAILED AND TWO-TAILED TESTS

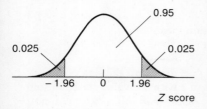

Fig. 11.3   Two-tailed test

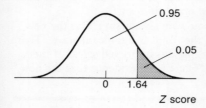

Fig. 11.4   One-tailed test

Table 11.1

Significance level	Two tailed	One tailed
5%	1.96	1.64
1%	2.58	2.33
0.2%	3.09	2.88

If a company were planning to introduce a new packaging for a product, it would probably test the new packaging in a sample of shops before deciding on the change. The hypothesis examined would be that the new packaging *sells more* than the old packaging. In this situation we are testing a change in one direction, so that we say we are conducting a *one-tailed test*. On the other hand, suppose we were conducting a survey of mothers of new-born babies, and wished to check whether the sample was representative of all mothers of new-born babies. In this case we would wish to test whether the mean age of the sample was *below or above* the mean age of all mothers. This would be a *two-tailed test* since the sample mean could be less than or greater than the population mean.

In examination questions you must first decide which sort of test to apply. If the question uses words which imply a change in *one direction* – words such as 'better', 'worse', 'improved', 'increased', 'reduced', etc. – then a *one-tailed test* must be employed. If the question implies that a change in *either direction* is important – perhaps asking 'is there any difference', 'is there any change' – then a *two-tailed test* must be used.

In using significance tests we have to find the *significant points*. We found in Chapter 10 that a z score of plus or minus 1.96 corresponded to 0.95 of the distribution being *within* these scores. This was on the basis that the area of both tails combined was 0.05 or 5%. Thus 1.96 is the figure if you are asked to use a *two-tailed* test at the 5% significance level. The corresponding figure for a *one-tailed test* at the 5% significance level is 1.64. Figure 11.3 shows the areas for a two-tailed test; Fig. 11.4 for a one-tailed test.

The *significant points* at 5%, 1% and 0.2% are given in Table 11.1 for two-tailed and one-tailed tests.

It is true you can obtain these results from the normal tables supplied in the examination room, but you can make mistakes in using the tables. It is strongly recommended that you **learn** these results, particularly the 5% and 1% points.

PROCEDURE IN CONDUCTING A SIGNIFICANCE TEST

There are five steps to follow; omission of any of these steps is likely to mean loss of marks. For illustration we shall use a one-tailed test:

(i)   Set out null hypothesis and alternative hypothesis. If you are testing for (say) an increase, this would mean:

Null hypothesis $\quad\quad\quad\quad \mu = \mu_0$
Alternative hypothesis $\quad\quad \mu > \mu_0$

(ii) State the level of significance for the test and give the corresponding significant point. In examination questions you are often told the level of significance: 1% or 5% are the most common. If you are not told, use the 5% level. If we are using a one-tailed test, this would mean:

One-tailed test 5% point: 1.64.

(iii) Calculate z using the formula:

$$z = \frac{\bar{x} - \mu_0}{s/\sqrt{n}},$$

$\bar{x}$ = sample mean
μ_0 = the hypothesised population mean
s = sample standard deviation
n = sample size

Note the *numerical* value of z, 'i.e. disregard the sign'.

(iv) Compare the calculated value of z with the value stated in step (ii). If the numerical value of z is *less than this value*, state 'not significant', and accept the null hypothesis. If the value of z is *greater than this value*, state 'significant at (say) 5% level', and *reject* the null hypothesis. If the value of z were also greater than (say) the 0.2% level, you could point this out.

(v) Conclusion. In the question you were asked (say) 'Has the new packaging improved sales?' If the result *is* significant state: 'There is evidence that the new packaging has improved sales.' If the result is not significant, state: 'There is no evidence that the new packaging has improved sales.'

Worked Example 3

A company is proposing to introduce a new system of production bonuses with the aim of improving productivity. Last year the average production per man per day was 1,020. Before introducing the bonuses throughout the company, it is decided to test the new bonus scheme on a random sample of 60 employees. The mean production per day for the sample was found to be 1,050 with a standard deviation of 120. Is there any evidence that the bonus scheme has improved productivity?

$\bar{x} = 1,050$, $\mu_0 = 1,020$, $s = 120$, $n = 60$

Null hypothesis $\quad\quad\quad\quad \mu = \mu_0 \quad\quad \mu = 1,020$
Alternative hypothesis $\quad\quad \mu > \mu_0 \quad\quad \mu > 1,020$

(The question asked 'improved productivity'; thus a one-tailed test is appropriate.)

One-tailed test 5% point: 1.64

$$z = \frac{\bar{x} - \mu_0}{s/\sqrt{n}} = \frac{1,050 - 1,020}{120/\sqrt{60}} = \frac{30}{15.49} = 1.94$$

$z = 1.94$ is greater than 5% point 1.64; therefore result is significant.

Reject null hypothesis. There is evidence that there has been an improvement in productivity.

| **Exercise 3** | A finance company is concerned about the arrears of payment on its loan accounts. In 1982 the average arrears of those accounts which were in arrears was £72.43. A survey conducted in 1983 using a simple random sample of 100 loan accounts which were in arrears showed that the mean arrears was £77.83 with a standard deviation of £16. Is there any evidence that there has been a significant change in the mean debt? |

(ICSA (part) June 1984)

(*Answer:* $z = 3.375$ – significant change)

SIGNIFICANCE TEST FOR A SINGLE PROPORTION

The procedure is very similar for that of a single mean. We use the fact that for large samples p the sample proportion is normally distributed. The hypotheses consider π_0 rather than μ_0. The formula for the calculation of z is:

$$z = \frac{p - \pi_0}{\sqrt{\dfrac{\pi_0(1 - \pi_0)}{n}}}$$

| **Worked Example 4** | Last year 15% of companies increased their staff. A survey of 1,000 companies showed that this year 20% of companies have taken on more staff. Is there any evidence that companies have recruited more staff? |

$p = 0.2$, $\pi_0 = 0.15$, $n = 1,000$

| Null hypothesis | $\pi = \pi_0$ | $\pi = 0.15$ |
| Alternative hypothesis | $\pi > \pi_0$ | $\pi > 0.15$ |

(The question asked 'more'; thus a one-tailed test should be used).

One-tailed test 5% point: 1.64.

$$z = \frac{p - \pi_0}{\sqrt{\dfrac{\pi_0(1 - \pi_0)}{n}}} = \frac{0.2 - 0.15}{\sqrt{\dfrac{0.15(1 - 0.15)}{1,000}}} = \frac{0.05}{0.0113} = 4.43$$

$z = 4.43$ is greater than 5% point 1.64; therefore result significant. Reject null hypothesis, z is also greater than 0.2% point 2.88.

We conclude that there is very good evidence that more companies are taking on staff this year.

Note: in Worked Example 3 the conclusion was written slightly differently: in this example the words 'very good' were included; this was because the result was significant at 0.2% and it was therefore *very unlikely* that the difference was due to the sample being unrepresentative.

Exercise 4

A sales department notices that 6% of interviews with sales prospects have matured into sales during 1978. In 1979 the sales department examined the results of a random sample of 2,000 interviews and noted 140 sales contracts were gained. Do you think that the proportion of successful interviews has significantly improved from 1978 to 1979?

(ICSA (part) June 1980)

(*Answer:* $z = 1.88$ – a significant improvement)

SIGNIFICANCE TEST FOR TWO MEANS

The procedure is similar to the procedure for that of a single mean. If the two sample means are $\bar{x}_1$ and $\bar{x}_2$, based on samples of sizes n_1 and n_2 respectively, the population means are μ_1 and μ_2 and the standard deviations are σ_1 and σ_2, it is possible to show that the difference between the two sample means, i.e. $\bar{x}_1 - \bar{x}_2$, is normally distributed with a mean $\mu_1 - \mu_2$ and standard error

$$\sqrt{\frac{\sigma_1{}^2}{n_1} + \frac{\sigma_2{}^2}{n_2}}.$$

The formula for the calculation of z is:

$$z = \frac{\bar{x}_1 - \bar{x}_2}{\sqrt{\dfrac{s_1^2}{n_1} + \dfrac{s_2^2}{n_1}}},$$

where s_1 and s_2 are sample estimates of σ_1 and σ_2.

Worked Example 5

A survey based on simple random samples was conducted to examine the income of women manual workers in the south-east and south-west regions of England. The survey showed that in the south-east the sample involved 256 women, the mean weekly income was £72.58 with a standard deviation of £15.20. For the south-west the corresponding figures were 170, £68.90, and £13.40 respectively. Is there any evidence at the 1% significance level of a difference between the income levels in the two regions?

$\bar{x}_1 = 72.58$, $\bar{x}_2 = 68.90$, $n_1 = 256$, $n_2 = 170$, $s_1 = 15.20$, $s_2 = 13.40$

Null hypothesis $\qquad\qquad \mu_1 = \mu_2$

Alternative hypothesis $\qquad \mu_1 \neq \mu_2$

(The question asked 'difference'; thus two-tailed test.)

Two-tailed test 1% point: 2.58

$$z = \frac{\bar{x}_1 - \bar{x}_2}{\sqrt{\dfrac{s_1^2}{n_1} + \dfrac{s_2^2}{n_2}}} = \frac{72.58 - 68.90}{\sqrt{\dfrac{15.20^2}{256} + \dfrac{13.40^2}{170}}} = \frac{3.68}{1.3995} = 2.63$$

$z = 2.63$ is greater than 1% point 2.58; therefore reject null hypothesis.

There is evidence of a significant difference at the 1% level of income in the two regions.

(a) Distinguish between a one-tailed and a two-tailed test.

(b) A company which manufactures pocket calculators has invited tenders for the supply of batteries. Two large, well-established rival firms have tendered, and samples of batteries from both of these have been tested. A sample of 150 batteries from the first supplier had a mean life of 1,643 hours with a standard deviation of 80 hours; a sample of 100 batteries from the second supplier had a mean life of 1,671 hours with a standard deviation of 93 hours.

Test the following hypotheses at a 0.01 level of significance:

(i) that the difference in the mean lives of the batteries is significant;

(ii) that the batteries from the second supplier last longer than those from the first supplier.

(ABE June 1983)

(*Answers:* $z = 2.46$; (i) not significant; (ii) significant)

SIGNIFICANCE TEST OF TWO PROPORTIONS

The procedure is similar to that of testing a single proportion. If the two sample proportions are p_1 and p_2, based on samples of size n_1 and n_2 respectively, and the population proportions are π_1 and π_2, it is possible to show that the difference between the two sample proportions, i.e. $p_1 - p_2$, is normally distributed with mean $\pi_1 - \pi_2$ and standard error given by the following formula:

$$\sqrt{p(1-p)\left(\frac{1}{n_1} + \frac{1}{n_2}\right)} \text{ where } p = \frac{n_1 p_1 + n_2 p_2}{n_1 + n_2}$$

Note that in some textbooks a different formula is given:

$$\sqrt{\frac{p_1(1-p_1)}{n_1} + \frac{p_2(1-p_2)}{n_2}}$$

Most examiners do not give full marks for the use of the second formula.

The formula for z is:

$$z = \frac{p_1 - p_2}{\sqrt{p(1-p)\left(\frac{1}{n_1} + \frac{1}{n_2}\right)}} \text{ where } p = \frac{n_1 p_1 + n_2 p_2}{n_1 + n_2}$$

Worked Example 6

A company manufactures a component using an existing machine. It needs to expand production and buys a second-hand machine. To test the effectiveness of the additional machine, random samples are taken from the output of both machines and the number of defectives produced is noted. The results for the old machine were: sample size 150, number of defectives 18; the corresponding results for the additional machine were 125, 20. Is there any difference between the two machines?

$$n_1 = 150, \ p_1 = \frac{18}{150} = 0.12, \ n_2 = 125, \ p_2 = \frac{20}{125} = 0.16.$$

Null hypothesis $\qquad\qquad \pi_1 = \pi_2$

Alternative hypothesis $\qquad \pi_1 \neq \pi_2$

(The question asked 'difference'; thus two-tailed test.)

Two-tailed test 5% point: 1.96

$$z = \frac{p_1 - p_2}{\sqrt{p(1-p)\left(\dfrac{1}{n_1} + \dfrac{1}{n_2}\right)}} \text{ where } p = \frac{n_1 p_1 + n_2 p_2}{n_1 + n_2}$$

$$p = \frac{150 \times 0.12 + 125 \times 0.16}{150 + 125} = \frac{38}{275} = 0.1382$$

$$z = \frac{0.12 - 0.16}{\sqrt{0.1382(1 - 0.1382)\left(\dfrac{1}{150} + \dfrac{1}{125}\right)}} = \frac{-0.04}{0.04179} = -0.96$$

we take the numerical value of $z = 0.96$.

$z = 0.96$ is less than 5% point 1.96; therefore accept null hypothesis. There is no evidence of a difference between the two machines.

Exercise 6

A firm examined a random sample of 3,000 sales interviews in 1979, and of those, 175 were successful. Do you think that there is any significant difference between the proportions of successful interviews from this firm and for the one referred to in Exercise 4 during 1979?

(ICSA (part) June 1980)

(*Answer:* $z = 1.66$ – no significant difference)

SAMPLE SIZE TO OBTAIN STATED ACCURACY

Worked Example 7

An opinion poll is conducted based on 1,000 interviews; 45% of those interviewed state that they would vote for a particular party. Find a 95% confidence interval for the population voting for this party in the electorate. How large a sample is necessary to obtain the result within 1% at the 95% confidence level?

We can find a 95% confidence interval using the method of Worked Example 2:

$$0.45 - 1.96\sqrt{\frac{0.45(1 - 0.45)}{1,000}} \quad \text{to} \quad 0.45 + 1.96\sqrt{\frac{0.45(1 - 0.45)}{1,000}}$$

$$= 0.45 - 0.0308 \quad \text{to} \quad 0.45 + 0.0308 = 0.4192 \text{ to } 0.4808$$

$$= 41.92\% \text{ to } 48.08\%.$$

In the above result we were within 3.08%. We wish to reduce this to less than or equal to 1%.

We need to find n so that

$$1.96 \sqrt{\frac{0.45(1-0.45)}{n}} \leqslant 0.01$$

Squaring both sides,

$$3.8416 \times \frac{0.45(1-0.45)}{n} \leqslant 0.0001$$

Rearranging,

$$\frac{3.8416 \times 0.45(1-0.45)}{0.0001} \leqslant n$$

Thus $9507.96 \leqslant n$. Hence to obtain a result within 1% we require a sample of 9,508.

Exercise 7

An accountancy firm is investigating invoice errors in an audit. From a random sample of 400 invoices they found an error rate of 11%. Find a 95% confidence interval for the error rate. How large a sample would be needed to obtain the error rate within 2% at the 95% confidence level?

(*Answers:* 7.93% to 14.07%; 941)

C. SOLUTIONS TO EXERCISES

S1

(i) $175.72 - 1.96 \times \dfrac{42.6}{\sqrt{200}}$ to $175.72 - 1.96 \times \dfrac{42.6}{\sqrt{200}}$

$= 175.72 - 5.90$ to $175.72 + 5.90 = 169.82$ to 181.62.

(ii) $175.72 - 2.58 \times \dfrac{42.6}{\sqrt{200}}$ to $175.72 - 2.58 \times \dfrac{42.6}{\sqrt{200}}$

$= 175.72 - 7.77$ to $175.72 + 7.77 = 167.95$ to 183.49.

S2

$0.2 - 1.96 \sqrt{\dfrac{0.2(1-0.2)}{1,000}}$ to $0.2 + 1.96 \sqrt{\dfrac{0.2(1-0.2)}{1,000}}$

$= 0.2 - 0.0248$ to $0.2 + 0.0248 = 0.1752$ to $0.2248 = 17.52\%$ to 22.48%.
17.52% of 50,000 = 8,760; 22.48% of 50,000 = 11,240.

S3

Note: NH = null hypothesis; AH = alternative hypothesis.

NH: $\mu = 72.43$; AH: $\mu \neq 72.43$.

Two-tailed 5%: 1.96.

$$z = \frac{77.83 - 72.43}{16/\sqrt{100}} = \frac{5.40}{1.6} = 3.375;$$

$z > 1.96$ (and 3.09, the 0.2% point). Reject NH, conclude strong evidence that mean debt has changed.

S4

NH: $\pi = 0.06$; AH: $\pi > 0.06$.
One-tailed 5%: 1.64.

$$z = \frac{0.07 - 0.06}{\sqrt{\dfrac{0.06(1 - 0.06)}{2,000}}} = \frac{0.01}{0.00531} = 1.88;$$

$z > 1.64$. Reject NH, conclude a significant improvement in number of successful interviews.

S5

(a) Read the section on one-tailed and two-tailed tests; make the point that if change is in one direction, a one-tailed test is required. Sketches such as Figs 11.3 and 11.4 help.

(b) (i) NH: $\mu_1 = \mu_2$; AH: $\mu_1 \neq \mu_2$; two-tailed 0.01($= 1\%$): 2.58.

$$z = \frac{1,643 - 1,671}{\sqrt{\dfrac{80^2}{150} + \dfrac{93^2}{100}}} = \frac{-28}{11.36} = -2.46;$$

numerical value $+2.46$. z less than 2.58; accept NH conclude difference in lives of batteries not significant.

(ii) NH: $\mu_1 = \mu_2$; AH: $\mu_1 < \mu_2$, one-tailed 1% $= 2.33$; $z = 2.46$ (same as (i)); z greater than 2.33. Reject NH, conclude that batteries from second supplier last longer.

S6

NH: $\pi_1 = \pi_2$; AH: $\pi_1 \neq \pi_2$; two-tailed test 5%: 1.96.

$$p = \frac{175 + 140}{3,000 + 2,000} = 0.063,$$

$$z = \frac{0.05833 - 0.07}{\sqrt{0.063(1 - 0.063)\left(\dfrac{1}{3,000} + \dfrac{1}{2,000}\right)}} = \frac{-0.01167}{0.007014} = -1.66$$

$z = 1.66$ (numerical value), accept NH. No evidence of difference between proportions.

$$0.11 - 1.96 \sqrt{\frac{0.11(1-0.11)}{400}} \quad \text{to} \quad 0.11 + 1.96 \sqrt{\frac{0.11(1-0.11)}{400}}$$

$0.11 - 0.03066$ to $0.11 + 0.03066 = 0.07934$ to $0.14066 = 7.93\%$ to 14.07%

$$1.96 \sqrt{\frac{0.11(1-0.11)}{n}} < 0.02, \; 3.8416 \times \frac{0.11(1-0.11)}{n} < 0.0004$$

$$\frac{3.8416 \times 0.11(1-0.11)}{0.0004} < n, \; 940.23 < n, \; n = 941$$

D. RECENT EXAMINATION QUESTIONS

Q1

(a) Explain what is meant by the term 95% confidence interval.

(b) A random sample of 50 small businesses was selected recently and the management were asked if they expected to take on more staff over the next year. 20 replied that they did expect to.

Set up a 95% confidence interval for the proportion of small businesses expecting to take on more staff over the next year.

(c) In a random sample of 40 medium businesses, 7 expected to take on more staff over the next year.

Test whether there is a significant difference between small and medium businesses in the proportion expecting to take on more staff over the next year.

(LCCI April 1985)

(*Answers:* 0.2642 to 0.5358; $z = 2.31$ – a significant difference)

Q2

(a) Two factories of the same company each make supplies of a standard item; each was asked to provide a random sample of 1,000 items for testing. In the case of the first factory, the random sample of 1,000 items contained 20 defectives; the second factory's random sample of 1,000 items contained 25 defectives. Is there any significant difference between the defectiveness of the two random samples?

(b) Random samples of 100 items were selected from the production lines in a quality test; the first test related to weights, and it was found that the mean weight of the first 100 items was 450 grams (the standard deviation was 2 grams). The second random sample of 100 items provided a mean weight of 452 grams (again, the standard deviation was 2 grams). Is there any significant difference between the mean weights of items in the two random samples?

(ICSA (part) June 1982)

Q3

A company takes a random sample of 200 orders from the order book. The value of the orders was distributed as shown in Table Q3.

Table Q3

Value of order	Percentage of orders
£100 and under £200	7
£200 and under £300	13
£300 and under £400	35
£400 and under £500	19
£500 and under £700	16
£700 and under £900	10

(i) Calculate the mean and standard deviation of the value of the order.
(ii) Find a 95% confidence interval for the mean value of the order.

(ICSA Dec. 1982)

(*Answers:* £427; £174.1; £402.87 to £451.13)

Q4

(a) A printing firm which manufactures coloured labels for the packing trades has found over past years that the fraction defective on its main production line has been 3%. A random sample of coloured labels containing 500 items has just been checked and is found to have 25 defective labels. Is this evidence of a significant increase in defectiveness?

(b) A random sample of 1,000 labels from a different firm has been found to contain 60 defective items – is this sample significantly different in its level of defectiveness from the sample of 500 items with 25 defectives?

(ICSA Dec. 1979)

(*Answers:* z = 2.62 – a significant increase; z = 0.79 – not significantly different)

Q5

A company decided to examine bad debts. A random sample of 200 bad debts was taken; the distribution of the length of life of these bad debts is given in Table Q5.

Table Q5

Number of working days	Percentage of bad debts
1–5	22
6–10	25
11–15	21
16–20	14
21–25	8
26–30	7
31–35	3

(i) Calculate the mean and standard deviation of length of life of bad debts.
(ii) In the previous year and mean length of life of bad debts was 11.4 working days. Is there any evidence that the mean length of life of bad debts has changed?

(ICSA June 1983)

(*Answers:* 12.7; 8.21; z = 2.24 – evidence of a change)

E. OUTLINE ANSWERS TO EXAM QUESTIONS

A1

(a) See section on confidence intervals.

(b) $0.4 - 1.96 \sqrt{\dfrac{0.4(1-0.4)}{50}}$ to $0.4 + 1.96 \sqrt{\dfrac{0.4(1-0.4)}{50}}$

$= 0.4 - 0.1358$ to $0.4 + 0.1358 = 0.2642$ to 0.5358.

(c) NH: $\pi_1 = \pi_2$; AH: $\pi_1 \neq \pi_2$; two-tailed test 5%: 1.96.

$p = \dfrac{20 + 7}{50 + 40} = 0.3$, $z = \dfrac{0.4 - 0.175}{\sqrt{0.3(1-0.3)\left(\dfrac{1}{50} + \dfrac{1}{40}\right)}} = \dfrac{0.225}{0.0972} = 2.31$;

z greater than 1.96. Reject NH, conclude that there is evidence of a significant difference in proportion expecting to take on more staff.

A2

(a) NH: $\pi_1 = \pi_2$; AH: $\pi_1 \neq \pi_2$; two-tailed test 5%: 1.96.

$p = \dfrac{20 + 25}{1,000 + 1,000} = 0.0225$,

$z = \dfrac{0.02 - 0.025}{\sqrt{0.0225(1-0.0225)\left(\dfrac{1}{1,000} + \dfrac{1}{1,000}\right)}} = \dfrac{0.005}{0.006632} = 0.75$

z less than 1.96; accept NH, no evidence of a difference of defectiveness of two factories.

(b) NH: $\mu_1 = \mu_2$; AH: $\mu_1 \neq \mu_2$; two-tailed test 5%: 1.96.

$z = \dfrac{452 - 450}{\sqrt{\dfrac{2^2}{100} + \dfrac{2^2}{100}}} = \dfrac{2}{0.2828} = 7.07$;

z greater than 1.96 (and also 3.09, the 0.2% point). Reject NH, very strong evidence of a significant difference.

A3

(i) $\Sigma f = 100$, $\Sigma fx = 42,700$, $\Sigma fx^2 = 21,265,000$,

$\bar{x} = \dfrac{\Sigma fx}{\Sigma f} = \dfrac{42,700}{100} = 427$

$s = \sqrt{\dfrac{\Sigma fx^2}{\Sigma f} - \left(\dfrac{\Sigma fx}{\Sigma f}\right)^2} = \sqrt{\dfrac{21,265,000}{100} - \left(\dfrac{42,700}{100}\right)^2}$
$= \sqrt{212,650 - 182,329} = 174.1$

(ii) $427 - 1.96 \times \dfrac{174.1}{\sqrt{200}}$ to $427 + 1.96 \times \dfrac{174.1}{\sqrt{200}}$

$= 427 - 24.13$ to $427 + 24.13 = £402.87$ to $£451.13.$

A4

(a) NH: $\pi = 0.03$; AH: $\pi > 0.03$; one-tailed test 5%: 1.64.

$$z = \frac{0.05 - 0.03}{\sqrt{\dfrac{0.03(1 - 0.03)}{500}}} = \frac{0.02}{0.00763} = 2.62;$$

z greater than 1.64 (and 2.33, the 1% point). Reject NH, conclude good evidence of increase in defectiveness.

(b) NH: $\pi_1 = \pi_2$; AH: $\pi_1 \neq \pi_2$; two-tailed test 5%: 1.96.

$$p = \frac{25 + 60}{500 + 1,000} = 0.05667,$$

$$z = \frac{0.06 - 0.05}{\sqrt{0.05667(1 - 0.05667)\left(\dfrac{1}{1,000} + \dfrac{1}{500}\right)}} = \frac{0.01}{0.01266}$$

$= 0.79$ z less than 1.96, accept NH, no evidence of a significant difference between firms.

A5

(i) $\Sigma f = 100$, $\Sigma fx = 1,270$, $\Sigma fx^2 = 22,870$,

$$\bar{x} = \frac{\Sigma fx}{\Sigma f} = \frac{1,270}{100} = 12.7$$

$$s = \sqrt{\frac{\Sigma fx^2}{\Sigma f} - \left(\frac{\Sigma fx}{\Sigma f}\right)^2} = \sqrt{\frac{22,870}{100} - \left(\frac{1,270}{100}\right)^2} = \sqrt{228.7 - 161.29}$$
$$= 8.21$$

(ii) NH: $\mu = 11.4$; AH: $\mu \neq 11.4$; two-tailed test 5%: 1.96.

$$z = \frac{12.7 - 11.4}{\dfrac{8.21}{\sqrt{200}}} = \frac{1.3}{0.5805} = 2.24;$$

z greater than 1.96. Reject NH, evidence that length of life of bad debts has changed.

A STEP FURTHER

Mulholland and Jones, *Fundamentals of Statistics*, Ch. 9.

Chapter 12

Algebra applied to business and economics

A. GETTING STARTED

In Chapter 3 we reviewed the basic rules of algebra. A number of important techniques in business and economics rely on the application of these rules. *Break-even analysis* is such a technique, being used to find the level of output at which total revenue equals total cost. The firm must be able to produce and sell this *minimum* level of output if it is to have a long-term future. Algebra can be usefully applied to *revenue* and *cost* functions. It can also be used to derive the *equilibrium* level of output at which demand equals supply, and to calculate the price at which that output can be sold.

In a number of courses *matrix algebra* is part of the Syllabus. A section of this chapter reviews matrix algebra and applies the technique to a number of problems which are frequently examined. If your syllabus excludes matrix algebra you can ignore this section.

B. ESSENTIAL PRINCIPLES

BREAK-EVEN ANALYSIS

This technique is used in accounting. The aim is to find the 'break-even' point, i.e. the level of activity where the total revenue for a period will equal the total costs for the same period. Suppose that q is the output in the given period and that this output is sold at a constant price p. The revenue R is given by quantity sold multiplied by the price:

$$R = pq$$

In making a product, we incur two types of cost:

(a) *fixed costs* – those costs involved in the absence of any production; costs of buildings and equipment, administrative salaries, etc.
(b) *variable costs* – those costs involved in the actual manufacturing process; materials, fuel, operative wages, etc.

If F denotes fixed costs and v denotes variable costs per unit of output, then to make a quantity q of the product the *variable costs* would be $v \times q$. The *total costs* C would be given by the following expression:

total costs = fixed costs + variable costs

$$C = F + vq$$

Now the 'break-even' point is when total revenue equals total costs i.e.

$$R = C$$

Hence $pq = F + vq$

We need to find q; if we are to find q using *algebra* we need to make q the subject of the formula:

$$pq - vq = F \qquad (p-v)q = F \qquad q = \frac{F}{(p-v)}$$

The break-even point can also be found *graphically*. The graphical method is explained in Worked Example 1.

Worked Example 1

A firm makes a cabinet. The selling price is £60. The variable costs per cabinet are £20. The fixed costs for the period is £12,000. Find the break-even point. Find also the level of output for which the profit is £2,000.

$$R = pq \qquad R = 60q$$
$$C = F + vq \qquad C = 12,000 + 20q$$

The lines $R = 60q$ and $C = 12,000 + 20q$ are plotted in Fig. 12.1.

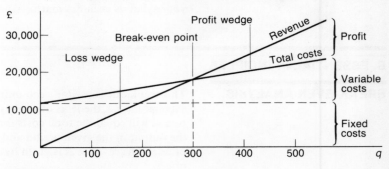

Fig. 12.1

From the graph we see that the break-even point is 300 cabinets. An output of less than 300 will involve a loss; the amount of the loss is shown by the 'loss wedge'. An output of more than 300 will involve a profit; the amount of the profit is shown by the 'profit wedge'.

We may also obtain the result algebraically.

At the break-even point revenue equals costs, i.e. $R = C$

$60q = 12,000 + 20q$ ∴ $60q - 20q = 12,000$ ∴ $40q = 12,000$ ∴ $q = 300$

We can also use the formula already derived:

$$q = \frac{F}{p - v} = \frac{12,000}{60 - 20} = \frac{12,000}{40} = 300$$

If you are asked to draw a break-even chart, it is a good idea to solve it algebraically first, so that you can work out the scale for the chart. If you find that the break-even point is (say) 500, then you know that the horizontal scale must go beyond 500.

We can also use the technique of break-even analysis to find the level of output which yields *specific levels of profit*. For instance, to find the level of output for which the profit is £2,000, we note that profits equal revenue minus costs. If P is profits then

$P = R - C$

$2,000 = 60q - (12,000 + 20q)$ ∴ $2,000 = 60q - 12,000 - 20q$
∴ $14,000 = 40q$ ∴ $q = 350$

Exercise 1

A company makes a product: the variable costs per item are £10, the fixed costs are £900. If the selling price is £25 find (a) the break-even point, (b) the output level which produces a profit of £300. (*Answers:* 60; 80)

A related problem in this area examines the total costs for different production methods; the problem and its solution are covered in Worked Example 2.

Worked Example 2

A company has two possible production methods: (a) capital intensive where the fixed costs are £30,000 and the variable costs are £10 per unit; (b) labour intensive where the fixed costs are £10,000 and the variable costs are £15 per unit. Find the level of output for which the total costs are the same.

Let q be the level of output. In case (a) total costs are given by:

$C_a = 30,000 + 10q$
In case (b) the total costs are:
$C_b = 10,000 + 15q$
When the total costs are the same, $C_a = C_b$
$30,000 + 10q = 10,000 + 15q$ ∴ $20,000 = 5q$ ∴ $q = 4,000$

The company should use method (b) if output is below 4,000, method (a) if output is above 4,000.

Exercise 2

A firm has two possible production processes for a new product. For process A the fixed costs per year are £20,000 and the variable costs per unit are £5. For process B the corresponding costs are £12,000 and £10 respectively. Find the level of output for which the total costs are the same.

If the selling price is £15 irrespective of the process used, find the break-even point for each process.

(*Answers:* 1,600; 2,000; 2,400)

Worked Example 3

A company wishes to hire a lorry for a week (the company works a five-day week). Three lorry hire firms were approached, and these firms submitted the following tariffs:

　　Firm A – £20 per day plus 50 pence per mile.
　　Firm B – £30 per day plus 25 pence per mile.
　　Firm C – £250 per week with no mileage charge.

(i)　Show on the same graph how the three tariffs vary.
(ii)　Advise the company.

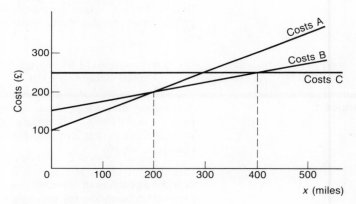

Fig. 12.2

Let x be the number of miles covered each week. For firm A the mileage charge would be £0.5x, the weekly charge would be 5×20 = £100. Thus the total charge would be:

$$C_A = 100 + 0.5x$$

The corresponding results for firms B and C are:

$$C_B = 150 + 0.25x$$
$$C_C = 250$$

These results are plotted in Fig. 12.2.

For this problem there are two break-even points – at 200 and at 400 miles. Assuming that the aim of the company is to minimize hiring costs, then the company should employ firm A if the mileage is less than 200 miles; firm B if the mileage is between 200 and 400 miles; and firm C if over 400 miles.

Exercise 3

A sales representative wishes to hire a car while his own car is being repaired. He approaches a car hire firm who quote three different tariffs:

Tariff A – a fixed charge of £80 per week.
Tariff B – a charge of 30 pence per mile.
Tariff C – a fixed charge of £20 plus 20 pence per mile.
Advise the sales representative.

(*Answer:* less than 200, B; 200–300, C; over 300, A)

REVENUE AND COST FUNCTIONS

In the previous section we met *simple* revenue and cost functions:

$$R = pq \qquad C = F + vq$$

These functions when plotted produced *straight lines*; in practice these functions could be *curves*.

The *revenue function* assumed that the price was fixed. There is a case for arguing that as the quantity increases, the price will fall. For instance, the relation between price and quantity could be:

$$p = 12 - q$$

In this situation the revenue function would be a curve and *not* a straight line. Revenue is price multiplied by quantity, i.e.

$$R = pq, \qquad R = (12 - q)q, \qquad R = 12q - q^2$$

The plotting of this *curve* was set in Exercise 9 of Chapter 3 (see Fig. 3S.4).

The *cost function* $C = F + vq$ assumes that the variable costs are constant for all levels of output. It is likely that, when the output rises above a certain level, we may have to start paying workers at overtime rates, so that variable costs would increase. Such a cost function could be:

$$C = 10 + 2q + q^2$$

Which again would be a curve and *not* a straight line.

Worked Example 4

Plot the cost function $C = 10 + 2q + q^2$.

Table 12.1

q	0	1	2	3	4	5
q^2	0	1	4	9	16	25
$2q$	0	2	4	6	8	10
10	10	10	10	10	10	10
$C = 10 + 2q + q^2$	10	13	18	25	34	45

Using the calculations of Table 12.1, the cost function is plotted in Fig. 12.3.

179

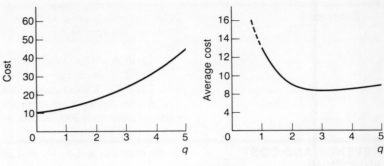

Fig. 12.3   Fig. 12.4

Exercise 4   Plot the cost function $C = 2q^2 + q + 5$.

AVERAGE COST FUNCTION

In economics we are often interested in the average cost. If we know the cost function, it is easy to find and to plot the *average cost function*:

$$\text{Average cost} = \frac{\text{total cost}}{\text{quantity}} = \frac{C}{q}$$

Worked Example 5   If $C = 10 + 2q + q^2$ find an expression for average cost and plot the average cost function.

$$\text{Average cost} = \frac{C}{q} = \frac{10 + 2q + q^2}{q} = \frac{10}{q} + 2 + q.$$

The calculations for this function are shown in Table 12.2.

Table 12.2

q	1	2	3	4	5
q	1	2	3	4	5
2	2	2	2	2	2
$\dfrac{10}{q}$	10	5	3.33	2.5	2
$\dfrac{10}{q} + 2 + q$	13	9	8.33	8.5	9

The average cost function is plotted in Fig. 12.4.

Note: When $q = 0$, $10/q$ is infinitely large, so we do not include $q = 0$ in Table 12.2. However, if you are asked to plot from $q = 0$ to $q = 5$, then the part for q less than 1 is shown as the dotted line in Fig. 12.4.

You can also be asked to plot the *average variable cost function*. In the expression $C = 10 + 2q + q^2$, 10 is the *total fixed cost*, and $2q + q^2$ is the *total variable cost*. The *average variable cost* would therefore be $2 + q$.

Exercise 5

If $C = 5 + 2q + q^2$, find the average cost and plot the average cost function.

SUPPLY AND DEMAND EQUATIONS

Let us consider a situation where the supply and demand equations are straight lines. The equilibrium position occurs when supply equals demand. In mathematical terms this can be expressed as:

$$q^s = ap + b \qquad \ldots (1)$$
$$q^d = cp + d \qquad \ldots (2)$$
$$q^s = q^d \qquad \ldots (3)$$

Equations (1) and (2) represent the supply and demand for a commodity in relation to price. Equation (3) states that the *equilibrium price* is that price which equates supply and demand.

In Chapter 6 we argued that the *dependent* variable should be placed on the vertical axis. If we plotted supply and demand equations we would, therefore, expect to place q on the vertical axis, since both supply and demand *depend* on price. In economics textbooks, however, q is placed on the horizontal axis. This is such an established practice that we shall continue to follow it. In Fig. 12.5 the lines are plotted in the same way as in economics books.

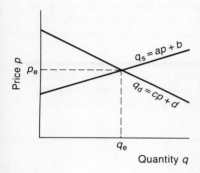

Fig. 12.5

Worked Example 6

We are given the following supply and demand equations:

$$q_s = 2p - 100 \qquad \ldots (1)$$
$$q_d = -0.5p + 500 \qquad \ldots (2)$$

Find (i) graphically; (ii) algebraically, the level of output and the corresponding price when $q_s = q_d$.

(i) From (1), if $q = 0$, $p = 50$; if $q = 500$, $p = 300$.
   From (2), if $q = 0$, $p = 1,000$; if $q = 500$, $p = 0$.
These lines are plotted in Fig. 12.6. The equilibrium price is 240 and the quantity is 380.

(ii) At equilibrium $q_s = q_d$
$$2p - 100 = -0.5p + 500$$
$$2p + 0.5p = 100 + 500$$
$$2.5p = 600$$
$$p = 240$$
To find q, substitute in equation (1):
$$q_s = 2 \times 240 - 100 = 380$$
As a check, substitute in equation (2):
$$q_d = -0.5 \times 240 + 500 = 380$$

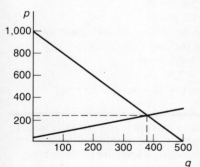

Fig. 12.6

(a) Find the equilibrium price and quantity for the following:
  (i) $q_s = 35 - p$ (ii) $q_s = 30 - 2p$
    $q_d = -7 + 0.5p$ $q_d = -12 + 1.5p$
(b) If the demand curve is $p_d = q^2 - 20q + 200$ and the supply curve is $p_s = 350 + 5q$, find the equilibrium price and quantity.
(*Answers:* 28, 7; 12, 6; 500, 30)

OBTAINING THE EQUATIONS OF SUPPLY AND DEMAND CURVES

How to obtain the equations of supply and demand curves in practice from empirical data is outside the scope of this book. Examination questions set at the introductory and foundation level are based on either regression lines (see Chapter 6) or the method shown in Worked Example 7.

Worked Example 7

When the output is 10 units a company would have to charge a price of £20 per unit; for an output of 50 units the corresponding price would be £15. Assuming a linear relationship between price and quantity, find the supply equation.

Let the equation of the supply curve be $q_s = a + bp$; we need to find the values of a and b
  When $q = 10$, $p = 20$ $10 = a + 20b$ ... (1)
  When $q = 50$, $p = 15$ $50 = a + 15b$ ... (2)
We now have two simultaneous equations. We solve using the methods of Chapter 3. Subtract equation (2) from equation (1):
  $-40 = 5b$ $b = -8$
  Substitute $b = -8$ in equation (1):
  $10 = a + 20 \times (-8)$ $10 = a - 160$ $a = 170$
Hence supply curve is: $q_s = 170 - 8p$

Exercise 7

(a) Assuming a linear relationship, find the supply equation if:
  (i) $p = 20$, $q = 10$ (ii) $q = 50$, $p = 15$
    $p = 10$, $q = 40$ $q = 10$, $p = 25$
(b) Assuming a linear relationship between costs and quantity produced, find the equation of the cost function if:
  (i) $C = 100$, $q = 20$ (ii) $C = 80$, $q = 0$
    $C = 300$, $q = 70$ $C = 220$, $q = 35$
(*Answers:* $q_s = 70 - 3p$; $q_s = 110 - 4p$; $C = 20 + 4q$; $C = 80 + 4q$)

MATRIX ALGEBRA APPLIED TO BUSINESS

You should first check whether matrix algebra is in your syllabus. If matrix algebra is excluded, then ignore this section. Questions set on matrices at the introductory and foundation level are of three types:

1. Inverting a matrix and/or solving simultaneous equations using matrices.
2. Forming matrices/vectors from the information supplied and then manipulating (usually multiplying) these matrices/vectors.
3. Brand switching.

In Chapter 3, type 1 was covered. The manipulation of matrices and/or vectors was also covered, although the actual *formation* of a matrix was not specifically covered. Forming a matrix does not usually present a problem to students, and a question of this type is set at the end of this chapter (see Exam Question Q8). This section will concentrate on type 3, i.e. using matrices in problems of brand switching.

BRAND SWITCHING

The technique described in this section is called *Markov analysis* and was developed to examine mathematically certain phenomena in science. Markov analysis has been used as a market research tool for examining and forecasting the behaviour of customers in response to advertising campaigns. Some customers will remain 'loyal' to a product; others will switch to other products as a result of the advertising campaign. The method of analysing 'brand switching' is explained in Worked Example 8.

Worked Example 8

Company A, which operates in the breakfast cereal market, is anxious to improve its share of the market. Company B and company C also serve the same market. Each company promotes its product by advertising. A survey is conducted to find out for company A and its competitors B and C, the number of customers who stay 'loyal' and those who switch to competitors. The survey shows that last month companies A, B and C had equal shares of the market, each selling 800. The gains and losses are shown in Table 12.3.

Table 12.3

Last Month's sales	Gains from			Losses to			Net Loss or gain	This month sales
	A	B	C	A	B	C		
A 800	—	240	80	—	160	240	80 (loss)	720
B 800	160	—	160	240	—	80	—	800
C 800	240	80	—	80	160	—	80	880

Assuming that total sales remain constant, and that the proportion changing also remains constant, what is likely to be: (i) next month's sales; (ii) the following month's sales; (iii) the equilibrium position in the long run?

From Table 12.3, $800 - (160 + 240) = 400$ of A's customers were 'loyal'; the corresponding numbers for B and C were 480 and 560 respectively. We can construct a table showing the proportion remaining 'loyal' and those switching (Table 12.4).

Table 12.4

		From				From	
	A	B	C		A	B	C
A	$\dfrac{400}{800}$	$\dfrac{240}{800}$	$\dfrac{80}{800}$	A	0.5	0.3	0.1
To B	$\dfrac{160}{800}$	$\dfrac{480}{800}$	$\dfrac{160}{800}$	To B	0.2	0.6	0.2
C	$\dfrac{240}{800}$	$\dfrac{80}{800}$	$\dfrac{560}{800}$	C	0.3	0.1	0.7

The table is called a *matrix of transition probabilities*. We can use this matrix to find this month's sales for the three companies. For company A this month's sales are $800 \times 0.5 + 800 \times 0.3 + 800 \times 0.1 = 720$. For company B, $800 \times 0.2 + 800 \times 0.6 + 800 \times 0.2 = 800$. For company C, $800 \times 0.3 + 800 \times 0.1 + 800 \times 0.7 = 880$. These calculations may be expressed in matrix form:

$$\begin{bmatrix} 0.5 & 0.3 & 0.1 \\ 0.2 & 0.6 & 0.2 \\ 0.3 & 0.1 & 0.7 \end{bmatrix} \begin{bmatrix} 800 \\ 800 \\ 800 \end{bmatrix} = \begin{bmatrix} 720 \\ 800 \\ 880 \end{bmatrix}$$

We can repeat this matrix multiplication to find next month's sales, assuming here that the proportion changing remains constant:

$$\begin{bmatrix} 0.5 & 0.3 & 0.1 \\ 0.2 & 0.6 & 0.2 \\ 0.3 & 0.1 & 0.7 \end{bmatrix} \begin{bmatrix} 720 \\ 800 \\ 880 \end{bmatrix} = \begin{bmatrix} 688 \\ 800 \\ 912 \end{bmatrix}$$

Thus the next month's sales are A 688, B 800, and C 912.

To find the following month's sales we repeat this process.

$$\begin{bmatrix} 0.5 & 0.3 & 0.1 \\ 0.2 & 0.6 & 0.2 \\ 0.3 & 0.1 & 0.7 \end{bmatrix} \begin{bmatrix} 688 \\ 800 \\ 912 \end{bmatrix} = \begin{bmatrix} 675.2 \\ 800 \\ 924.8 \end{bmatrix}$$

Thus the following month's sales are A 675.2, B 800, and C 924.8. We can see that company C is gaining from company A, company B is unchanged.

To find the equilibrium figure we could proceed as above, but the method is obviously very tedious. There is a method we can apply.

Let the equilibrium proportions be x, y and z. Clearly $x + y + z = 1$ and hence $z = 1 - x - y$.

In equilibrium, when we multiply by the transition matrix the proportion will be the same before *and* after multiplication. Thus

$$\begin{bmatrix} 0.5 & 0.3 & 0.1 \\ 0.2 & 0.6 & 0.2 \\ 0.3 & 0.1 & 0.7 \end{bmatrix} \begin{bmatrix} x \\ y \\ 1-x-y \end{bmatrix} = \begin{bmatrix} x \\ y \\ 1-x-y \end{bmatrix}$$

Multiply the matrix and the vector:

$$\begin{bmatrix} 0.5x+0.3y+0.1(1-x-y) \\ 0.2x+0.6y+0.2(1-x-y) \\ 0.3x+0.1y+0.7(1-x-y) \end{bmatrix} = \begin{bmatrix} x \\ y \\ 1-x-y \end{bmatrix}$$

If the two vectors are equal then their elements are equal; as we need to find x and y we need to take the first two elements only.

$$0.5x+0.3y+0.1(1-x-y)=x$$
$$0.2x+0.6y+0.2(1-x-y)=y$$

Multiplying by 10 to clear decimals,

$$5x+3y+(1-x-y)=10x$$
$$2x+6y+2(1-x-y)=10y$$
$$6x-2y=1$$
$$6y=2$$

Thus $y=0.3333$, $x=0.2778$ and $z=1-0.3333-0.2778=0.3889$. Hence the equilibrium share of the market for company A is $2{,}400 \times 0.2778 = 666.7$; the corresponding figures for companies B and C are 800 and 933.3 respectively.

It will be noted that, in the calculation of the equilibrium position, we made no use of the fact that initially the market shares were equal.

Exercise 8

Three dairies deliver milk in a town. At the beginning of the year dairy A has 40% of the market. Dairies B and C have 30% each. A market research survey shows that, during a period of a year, dairy A retains 90% of its customers and gains 5% of B's customers and 10% of C's customers. Dairy B retains 85% of its customers and gains 5% of A's customers and 15% of C's customers. Dairy C retains 75% of its customers and gains 5% of A's customers and 10% of B's customers.

(i) What will each dairy's share of the market be at the end of this year and the end of next year?

(ii) What will each dairy's share of the market be in equilibrium if this pattern of switching continues?

(*Answers:* 40.5, 32.0, 27.5; 40.80, 33.35, 25.85; 40.91, 36.36, 22.72)

C. SOLUTIONS TO EXERCISES

S1

$R=pq$, $R=25q$; $C=F+vq$, $C=900+10q$,
(a) $R=C$, $25q=900+10q$, $15q=900$, $q=60$
(b) $P=R-C$, $300=25q-(900+10q)$, $1200=15q$, $q=80$

S2

$C=F+vq$, $C_A=20{,}000+5q$, $C_B=12{,}000+10q$, $C_A=C_B$,
$20{,}000+5q=12{,}000+10q$, $8{,}000=5q$, $q=1{,}600$; $R=pq$, $R=15q$
$R=C$, $15q=20{,}000+5q$, $10q=20{,}000$, $q=2{,}000$; $15q=12{,}000+10q$,
$5q=12{,}000$, $q=2{,}400$.

S3

$C_A = 80$, $C_B = 0.3x$, $C_C = 20 + 0.2x$.

Figure 12S.1 shows that below 200 miles tariff B is preferable; between 200 and 300 miles tariff C; above 300 tariff A.

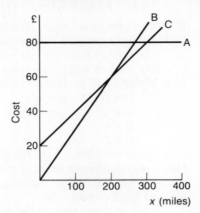

Fig. 12S.1

S4

The function is plotted in Fig. 12S.2.

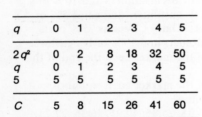

q	0	1	2	3	4	5
$2q^2$	0	2	8	18	32	50
q	0	1	2	3	4	5
5	5	5	5	5	5	5
C	5	8	15	26	41	60

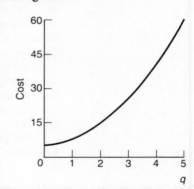

Fig. 12S.2

S5

$$A = \frac{C}{q} = \frac{5 + 2q^2 + q}{q}$$

$$= \frac{5}{q} + 2q + 1$$

The function is plotted in Fig. 12S.3.

q	1	2	3	4	5
$\dfrac{5}{q}$	5	2.5	1.67	1.25	1
$2q$	2	4	6	8	10
1	1	1	1	1	1
A	8	7.5	8.67	10.25	12

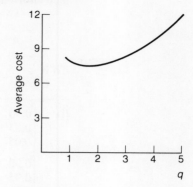

Fig. 12S.3

S6

(a) (i) $q_d = q_s, 35 - p = -7 + 0.5p, 42 = 1.5p, p = 28, q_2 = 35 - p,$
$q_d = 35 - 28, q_d = 7.$

    (ii) $30 - 2p = -12 + 1.5p, 42 = 3.5p, p = 12, q_d = 30 - 2p,$
$q_d = 30 - 2 \times 12 = 6.$

(b) $p_d = p_s, q^2 - 20q + 200 = 350 + 5q, q^2 - 25q - 150 = 0;$ this is a quadratic equation $(q - 30)(q + 5) = 0, q = 30$ (disregard $q = -5$ as negative quantities not possible). $p_s = 350 + 5q, p_s = 350 + 5 \times 30 = 500.$

S7

(a) (i) $q = a + bp$ $10 = a + 20b$ ...(1)
                            $40 = a + 10b$ ...(2)
Subtract (2) from (1): $-30 = 10b, b = -3$; substitute $b = -3$ in (1): $10 = a + 20 \times (-3), 10 = a - 60, a = 70$. Thus $q = 70 - 3p.$

    (ii) $q = a + bp$ $50 = a + 15b$ ...(1)
                            $10 = a + 25b$ ...(2)
Subtract (2) from (1): $40 = -10b, b = -4$; substitute $b = -4$ in (1): $50 = a + 15 \times (-4), 50 = a - 60, a = 110$. Thus $q = 110 - 4p.$

(b) (i) $C = a + bq$ $100 = a + 20b$ ...(1)
                            $300 = a + 70b$ ...(2)
Subtract (1) from (2): $200 = 50b, b = 4$; substitute $b = 4$ in (1): $100 = a + 20 \times 4, 100 = a + 80, a = 20$. Thus $C = 20 + 4q.$

    (ii) $C = a + bq$ $80 = a + 0b$ ...(1)
                            $220 = a + 35b$ ...(2)
From (1), $a = 80$; substitute $a = 80$ in (2): $220 = 80 + 35b,$
$35b = 140, b = 4$. Thus $C = 80 + 4q.$

The information may be placed into a matrix of percentages:

		From						From		
		A	B	C				A	B	C
	A	90	5	10			A	0.90	0.05	0.10
To	B	5	85	15	Change to matrix	To	B	0.05	0.85	0.15
	C	5	10	75	of probabilities		C	0.05	0.10	0.75

(i) $$\begin{bmatrix} 0.90 & 0.05 & 0.10 \\ 0.05 & 0.85 & 0.15 \\ 0.05 & 0.10 & 0.75 \end{bmatrix} \begin{bmatrix} 40 \\ 30 \\ 30 \end{bmatrix} = \begin{bmatrix} 40.5 \\ 32.0 \\ 27.5 \end{bmatrix}$$

$$\begin{bmatrix} 0.90 & 0.05 & 0.10 \\ 0.05 & 0.85 & 0.15 \\ 0.05 & 0.10 & 0.75 \end{bmatrix} \begin{bmatrix} 40.5 \\ 32.0 \\ 27.5 \end{bmatrix} = \begin{bmatrix} 40.80 \\ 33.35 \\ 25.85 \end{bmatrix}$$

(ii) $$\begin{bmatrix} 0.90 & 0.05 & 0.10 \\ 0.05 & 0.85 & 0.15 \\ 0.05 & 0.10 & 0.75 \end{bmatrix} \begin{bmatrix} x \\ y \\ 1-x-y \end{bmatrix} = \begin{bmatrix} x \\ y \\ 1-x-y \end{bmatrix}$$

$$\begin{bmatrix} 0.90x + 0.05y + 0.10(1-x-y) \\ 0.05x + 0.85y + 0.15(1-x-y) \\ 0.05x + 0.10y + 0.75(1-x-y) \end{bmatrix} = \begin{bmatrix} x \\ y \\ 1-x-y \end{bmatrix}$$

Comparing elements, $0.90x + 0.05y + 0.10(1-x-y) = x$
$0.05x + 0.85y + 0.15(1-x-y) = y$

Multiplying by 100 and rearranging, $10 = 20x + 5y$
$15 = 10x + 30y$

Solving, $x = 0.4091$, $y = 0.3636$, $z = 1-x-y = 0.2273$.
Percentages: A 40.91, B 36.36, C 22.72

D. RECENT EXAMINATION QUESTIONS

Q1

The marketing department estimates that if the selling price of a new product A1 is set at £40 per unit then the sales will be 400 units per week, while, if the selling price is set at £20 per unit, the sales will be 800 units per week. Assume that the graph of this function is linear.

The production department estimates that the variable costs will be £7.50 per unit and that the fixed costs will be £10,000 per week.
(a) Derive the cost, sales revenue, and profit functions.
(b) Graph the three equations derived in (a).
(c) From the graph estimate the maximum profit that can be obtained, stating the number of sales units and the selling price necessary to achieve this profit.

(ICMA May 1976)

(*Answers:* $C = 10{,}000 + 7.5q$, $R = 60q - 0.05q^2$, $P = -0.05q^2 + 52.5q - 10{,}000$; £3,781.25, 525, £33.75)

Q2

A manufacturer of fertilizers competes in a market in which price is not constant and his revenue function is not linear.

The manufacturer finds that his weekly production costs are £1,300 when producing 20 tonnes per week, and £1,700 when producing 30 tonnes per week. He finds that he can sell 20 tonnes per week at a price of £80 per tonne but has to reduce the price to £70 per tonne in order to sell 30 tonnes per week.

The manufacturer knows that the cost and price functions are linear. All production is sold.

(i)   Show that the weekly production costs, as a function of the quantity sold, are given by the equation $C = 500 + 40q$.

(ii) Find the price per tonne, p, as a function of the quantity sold.

(iii) Find the weekly revenue, R, as a function of the quantity sold.

(iv) At the break-even point, find the quantity sold, the price, and the weekly revenue.

(N.B. The break-even point is here defined as the smallest quantity sold at which total revenue equals total cost.)

(ICMA Nov. 1981)

(*Answers:* $P = 100 - q$; $R = 100q - q^2$; 10, £90, £900)

Q3

A ship travels between two ports. The cost of fuel is

$$£100(aX + \frac{b}{X} + 10)$$

where X is the average speed of the ship in knots ($X > 0$), and a and b are constants. If the ship travels at 4 knots the cost of fuel is £9,000, but at 6 knots it is £7,000. (1 knot = 1 nautical mile per hour.)

(a)   Find the values of a and b.

(b)   By a graphical method, or otherwise, find the speed giving optimum fuel economy and the cost of fuel at that speed.

(ICMA May 1984)

(*Answers:* 2; 288; 12; £5,800)

Q4

The total cost function is given by $C = x^2 + 16x + 39$, where x units is the quantity produced and £C the total cost.

(i)   Write down the expression for the average cost per unit.

(ii) Sketch the average cost function against x, for the values of x between $x = 0$ and $x = 8$.

(CACA (ACCA) (part) June 1984)

(*Answer:* $x + 16 + \frac{39}{x}$)

Q5

A company manufactures two products, A and B, by means of two processes: X and Y. The maximum capacity of process X is 1,900 hours and of process Y 5,000 hours. One unit of product A requires four hours in process X and two hours in process Y while one unit of product B requires one hour in process X and five hours in process Y.

You are required to calculate the number of units to be produced of products A and B to ensure that the maximum capacity available is utilized.

(ICMA (part) November 1979)

(*Answers:* 250; 900)

Q6

Last year a company began marketing a new (copyrighted) board game, Zqkoploy. Experience to date has shown that about 300 games per week can be sold at a selling price of £20 each, but if the price is reduced to £13, about 440 can be sold.

(a) Assuming a linear relationship, show that the approximate demand equation is:

$$P = 35 - \frac{1}{20} X$$

where P equals price per unit ($P > 0$), X equals quantity of units sold each week ($X > 0$).

(b) Write down the company's revenue (R) in terms of x.

(c) The company having fixed costs of £2,000 per week and variable costs of £$10X$ per week, write down the total cost (C) equation in terms of X.

(d) Determine at what point(s) the company breaks even.

(e) By drawing a graph or examining the break-even position or otherwise, advise the company on production and pricing policy for Zqkolopy if it wishes:
 (i) to maximize profit;
 (ii) to maximize revenue.

(ICMA (part) Nov. 1979)

(*Answers:* $R = 35X - \dfrac{X^2}{20}$; $C = 2,000 + 10X$; 100, 400; 250, £22.50; 350, £17.50)

Q7

The quantity of a commodity demanded or supplied is represented by q tonnes and £p represents its price. The demand equation is $2p - q = 15$ and the supply equation is $-p + 3q = 5$.

(i) Express these equations in matrix form.

(ii) Using matrix methods establish the values of p and q which satisfy the demand and supply equations simultaneously.

(CACA (ACCA) (part) Dec. 1984)

(*Answers:* $p = 10$; $q = 5$)

Q8

(a) Socsport plc is a company in the wholesale trade selling sportswear and stocks two brands, A and B, of football kit, each consisting of a shirt, a pair of shorts and a pair of socks.

 The costs for Brand A are £5.75 for a shirt, £3.99 for a pair of shorts and £1.85 for a pair of socks, and those for Brand B are

£6.25 for a shirt, £4.48 for a pair of shorts and £1.97 for a pair of socks. Three customers X, Y, Z demand the following combination of Brands; X, 36 kits of Brand A and 48 kits of Brand B; Y, 24 kits of Brand A and 72 kits of Brand B; Z, 60 kits of Brand A.

  (i)   Express the costs of Brands A and B in matrix form, then the demands of the customers, X, Y, Z in matrix form.

  (ii) By forming the product of the two matrices that you obtained in the previous part deduce the detailed costs to each of the customers.

(b) As a result of recent price increases Z has ceased to be a customer of Socsport but the demands of X and Y remain the same. Socsport no longer stocks socks of either brand. Re-express the demands of X and Y in matrix form and find its inverse matrix.

(CACA (ACCA) (part) June 1983)

(*Answers:* see solution)

E. OUTLINE ANSWERS TO EXAM QUESTIONS

A1

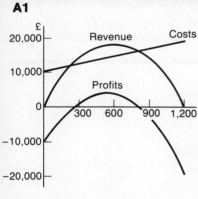

Fig. 12S.4

(a) $C = F + vq$, $C = 10{,}000 + 7.5q$

$R = pq$. We need an expression for p: $p = a + bq$, $40 = a + 400b$, $20 = a + 800b$. Solving, $b = -0.05$, $a = 60$, $p = 60 - 0.05q$, $R = (60 - 0.05q)q$, $R = 60q - 0.05q^2$

$P = R - C$, $P = 60q - 0.05q^2 - (10{,}000 + 7.5q)$, $P = -0.05q^2 + 52.5q - 10{,}000$.

(b) To work out scale note that $R = 0$ when $q = 0$ and $q = 1{,}200$. The graphs are shown in Fig. 12S.4.

q	0	200	400	600	800	1,000	1,200
C	10,000	11,500	13,000	14,500	16,000	17,500	19,000
R	0	10,000	16,000	18,000	16,000	10,000	0
P	-10,000	-1,500	3,000	3,500	0	-7,500	-19,000

(c) From the graph, maximum sales occur when $q = 525$. To find profit read off from graph or substitute $q = 525$ in equation for P: 3,781.25. To find price substitute $q = 525$ in equation for p: 33.75.

A2

  (i) $C = a + bq$, $1{,}300 = a + 20b$, $1{,}700 = a + 30b$. Solving, $b = 40$, $a = 500$; $C = 500 + 40q$.

  (ii) $p = a + bq$, $80 = a + 20b$, $70 = a + 30b$. Solving, $b = -1$, $a = 100$; $p = 100 - q$.

(iii) $R = pq$, $R = (100 - q)q$.

(iv) $R = C$, $(100 - q)q = (500 + 40q)$, $0 = q^2 - 60q + 500$.

Solving quadratic, $0 = (q - 10)(q - 50)$, $q = 10$ or $q = 50$. Question states smallest quantity, thus $q = 10$, $p = 100 - q$, $p = 100 - 10 = 90$; $R = (100 - q)q = (100 - 10) \times 10 = 900$.

A3

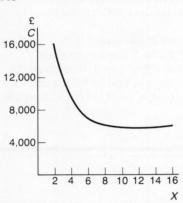

£

Fig. 12S.5

(a) $C = 100\left(aX + \dfrac{b}{X} + 10\right)$,

$9{,}000 = 100\left(4a + \dfrac{b}{4} + 10\right)$,

$7{,}000 = 100\left(6a + \dfrac{b}{6} + 10\right)$.

These equations simplify to $320 = 16a + b$ and $360 = 36a + b$. Solving, $a = 2$, $b = 288$.

(b) $C = 100\left(2X + \dfrac{288}{X} + 10\right)$.

X	2	4	6	8	10	12	14	16
C	15,800	9,000	7,000	6,200	5,880	5,800	5,857	6,000

From Fig. 12S.5 the minimum occurs when speed is 12 and cost £5,800. The 'otherwise' method is to use calculus (see Chapter 13).

A4

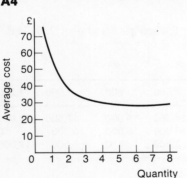

Fig. 12S.6

(i) $A = \dfrac{C}{x} = \dfrac{x^2 + 16x + 39}{x}$

$A = x + 16 + \dfrac{39}{x}$

The function is shown in Fig. 12S.6.

(ii)

x	0	1	2	3	4
A	∞	56.0	37.5	32.0	29.75
x	5	6	7	8	
A	28.8	28.5	28.57	28.88	

A5

Suppose A units of A and B units of B ensure maximum capacity. $1{,}900 = 4A + B$, $5{,}000 = 2A + 5B$. Solving, $A = 250$, $B = 900$.

(a) $P = a + bX$, $20 = a + 300b$, $13 = a + 440b$.

Solving, $b = \dfrac{-1}{20}$, $a = 35$; $P = 35 - \dfrac{1X}{20}$

(b) $R = PX$, $R = (35 - \dfrac{1X}{20})X$, $R = 35X - \dfrac{X^2}{20}$.

(c) $C = 2{,}000 + 10X$.

(d) $R = C$, $35X - \dfrac{1}{20}X^2 = 2{,}000 + 10X$;

simplifying, $0 = X^2 - 500X + 40{,}000$, $0 = (X - 100)(X - 400)$, $X = 100$, $X = 400$.

(e) Profit $= R - C = 35X - \dfrac{1}{20}X^2 - 2{,}000 - 10X$.

X	50	100	150	200	
P	-875	0	625	1,000	
R	1,625	3,000	4,125	5,000	

X	250	300	350	400	450
P	1,125	1,000	625	0	-875
R	5,625	6,000	6,125	6,000	5,625

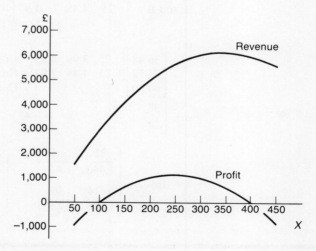

Fig. 12S.7

From Fig. 12S.7, maximum profit $X = 250$ and maximum revenue $X = 350$. The corresponding prices are found by substituting these values for X in equation

$$P = 35 - \frac{1}{20}X: \ P = 22.5 \text{ and } P = 17.5.$$

A7

(1)

$$\begin{matrix} 2p - q = 15 \\ -p + 3q = 5 \end{matrix} \qquad \begin{bmatrix} 2 & -1 \\ -1 & 3 \end{bmatrix} \begin{bmatrix} p \\ q \end{bmatrix} = \begin{bmatrix} 15 \\ 5 \end{bmatrix}$$

(ii) Find inverse of $\begin{bmatrix} 2 & -1 \\ -1 & 3 \end{bmatrix}$ $\qquad k = 2 \times 3 - (-1) \times (-1) = 5$

Inverse is $\begin{bmatrix} 0.6 & 0.2 \\ 0.2 & 0.4 \end{bmatrix}$ (see formula in Chapter 3)

$$\begin{bmatrix} 0.6 & 0.2 \\ 0.2 & 0.4 \end{bmatrix} \begin{bmatrix} 2 & -1 \\ -1 & 3 \end{bmatrix} \begin{bmatrix} p \\ q \end{bmatrix} = \begin{bmatrix} 0.6 & 0.2 \\ 0.2 & 0.4 \end{bmatrix} \begin{bmatrix} 15 \\ 5 \end{bmatrix}$$

$$\begin{bmatrix} 1 & 0 \\ 0 & 1 \end{bmatrix} \begin{bmatrix} p \\ q \end{bmatrix} = \begin{bmatrix} 10 \\ 5 \end{bmatrix} \qquad \begin{bmatrix} p \\ q \end{bmatrix} = \begin{bmatrix} 10 \\ 5 \end{bmatrix} \qquad \text{Thus } p = 10, \ q = 5.$$

A8

(a) (i)   Cost matrix

	Shirts	Shorts	Socks
Brand A	5.75	3.99	1.85
Brand B	6.25	4.48	1.97

Demand matrix

	Brand A	Brand B
Customer X	36	48
Customer Y	24	72
Customer Z	60	0

(ii)

$$\begin{bmatrix} 36 & 48 \\ 24 & 72 \\ 60 & 0 \end{bmatrix} \begin{bmatrix} 5.75 & 3.99 & 1.85 \\ 6.25 & 4.48 & 1.97 \end{bmatrix} =$$

	Shirts	Shorts	Socks
Customer X	507.00	358.68	161.16
Customer Y	588.00	418.32	186.24
Customer Z	345.00	239.40	111.00

(b) $\begin{bmatrix} 36 & 48 \\ 24 & 72 \end{bmatrix}$ $\qquad k = 72 \times 36 - 48 \times 24 = 1{,}440$

Inverse $\begin{bmatrix} \dfrac{72}{1{,}440} & \dfrac{-48}{1{,}440} \\ \dfrac{-24}{1{,}440} & \dfrac{36}{1{,}440} \end{bmatrix} = \begin{bmatrix} 0.05 & -0.0333 \\ -0.0167 & 0.025 \end{bmatrix}$

A STEP FURTHER   Tennant-Smith, *Mathematics for the Manager*, Chs 1, 2, 4, 5 and 6.

Chapter 13

Calculus

A. GETTING STARTED

You should first check whether this topic is included in your syllabus. At the time of writing both *differential* and *integral* calculus *are* included in the CACA (ACCA) syllabus. Although Calculus is not specifically included in the ICMA syllabus, ICMA questions such as those in Chapter 12 (Examination Questions 3 and 6) included the phrase 'by a graphical method or otherwise', indicating that a calculus method is acceptable. A calculus method is much quicker than a graphical method, so that ICMA students may consider studying this chapter up to page 203. Again many diploma and undergraduate courses which include options in quantitative methods would expect some familiarity with calculus and its applications.

Calculus in introductory and foundation level courses usually involves a practical application of calculus to business and economics. This chapter concentrates on the practical applications of calculus.

B. ESSENTIAL PRINCIPLES

MARGINAL COST

In economics *marginal cost* is defined as the change in cost for a unit change in quantity. In the case of the linear cost function $C = 100 + 5q$, if q increases by *one* unit, C increases by 5 units. This may be illustrated by finding C when $q = 10$ and 11 respectively; when $q = 10$, $C = 100 + 5 \times 10 = 150$; and when q increases by 1 unit, i.e. q becomes 11, $C = 100 + 5 \times 11 = 155$, showing that C has increased by $155 - 150 = 5$ units. Thus the marginal cost is equal to 5. The gradient of the cost function $C = 100 + 5q$ is also equal to 5. Hence the marginal cost is equal to the gradient of the cost function, i.e. *the rate of change of total cost*.

In the case of a cost function which is a *straight line* it is easy to find the gradient – it is the coefficient of q – which is therefore the marginal cost. In practice (and in examination questions!), however, cost functions are curves rather than straight lines. Figures 13.1 and 13.2 illustrate cost curves.

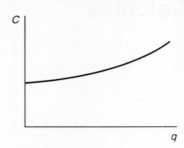

Fig. 13.1

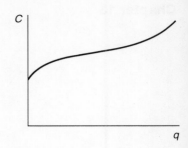

Fig. 13.2

In Figs. 13.1 and 13.2 the gradient of the *tangents* to the curves vary as q varies. As marginal cost is equal to the *gradient of the tangent*, we need to develop a technique which enables us to find the value of the gradient for various values of q.

DIFFERENTIAL CALCULUS

We can use differential calculus to find the gradient of a curve. In general, for a cost curve expressed by an equation involving cost C in terms of quantity q, the gradient of the curve is written

$$\frac{dC}{dq}.$$

We call $\frac{dC}{dq}$

the *differential coefficient*, and the process of finding it *differentiation*.

It is necessary to learn how to find $\frac{dC}{dq}$

for various expressions involving q. Calculus is often included in 'O'-level mathematics syllabuses, so the topic should be familiar to many.

However, at 'O'-level you are usually asked to find $\frac{dy}{dx}$. You have to remember that y is often replaced by C (for costs) or R (for revenue) or P (for profits), and that the x is often replaced by q (for quantity).

No attempt is made in this book to explain *how* the differential coefficients are found; at introductory and foundation level all that is necessary is that you learn the technique for finding differential coefficients. We start by looking at differential coefficients for cost curves where there are no fixed costs.

DIFFERENTIAL COEFFICIENTS

if $\quad C = q^2 \qquad \dfrac{\mathrm{d}C}{\mathrm{d}q} = 2q$

if $\quad C = q^3 \qquad \dfrac{\mathrm{d}C}{\mathrm{d}q} = 3q^2$

if $\quad C = q^4 \qquad \dfrac{\mathrm{d}C}{\mathrm{d}q} = 4q^3$

if $\quad C = q^5 \qquad \dfrac{\mathrm{d}C}{\mathrm{d}q} = 5q^4$

i.e. if $C = q^n \qquad \dfrac{\mathrm{d}C}{\mathrm{d}q} = nq^{n-1}$

So far the *constant term* associated with q has been the number 1, i.e. $1q^2$, $1q^3$, etc. In many cases the constant term may take on different values. For instance, to find $\dfrac{\mathrm{d}C}{\mathrm{d}q}$ if $C = 3q^2$, the answer is 3 times the differential coefficient of q^2, i.e.

$$\frac{\mathrm{d}C}{\mathrm{d}q} = 3 \times 2q = 6q.$$

In practice the easiest way to find such a differential coefficient is to multiply the constant term (3) by the power of q (2) which gives 6, and reduce the power of q by 1. So to find the differential coefficient of $6q^3$ multiply 6 and $3 = 18$ and reduce the power by 1, i.e. q^3 becomes q^2 and the result is $18q^2$. If we represent the constant term by the letter a we can derive the following result:

In general, if $C = aq^n \qquad \dfrac{\mathrm{d}C}{\mathrm{d}q} = anq^{n-1}$

if $C = 5q^3 \qquad \dfrac{\mathrm{d}C}{\mathrm{d}q} = 5 \times 3q^2 = 15q^2$

if $C = 7q^4 \qquad \dfrac{\mathrm{d}C}{\mathrm{d}q} = 7 \times 4q^3 = 28q^3$

You must learn the general result and be able to apply it.

It is important that you present a function in the *general form aq^n* before using the *general result*. The rules of powers (or indices) are helpful in getting a function into the general form. Read again p. 10 in Chapter 3.

In the expression $C = q^n$, the power n can take positive, negative or fractional values.

if $C = \dfrac{1}{q^2}$ then $C = q^{-2}$, i.e. $n = -2$

$$\frac{\mathrm{d}C}{\mathrm{d}q} = -2q^{-2-1} = -2q^{-3} = \frac{-2}{q^3}$$

$$\text{if } C = \sqrt{q} \text{ then } C = q^{\frac{1}{2}}, \text{ i.e. } n = \frac{1}{2}$$

$$\frac{dC}{dq} = \frac{1}{2}q^{\frac{1}{2}-1} = \frac{1}{2}q^{-\frac{1}{2}} = \frac{1}{2q^{\frac{1}{2}}} = \frac{1}{2\sqrt{q}}$$

$$\text{if } C = q \qquad \text{then } C = q^1, \text{ i.e. } n = 1$$

$$\frac{dC}{dq} = q^{1-1} = q^0 = 1$$

Exercise 1

Find $\dfrac{dC}{dq}$ if:

(i) $C = q^6$; (ii) $C = q^9$; (iii) $C = 4q^3$; (iv) $C = 5q^4$; (v) $C = 10q$;

(vi) $C = \dfrac{1}{q}$; (vii) $\dfrac{1}{\sqrt{q}}$

(*Answers:* $6q^5$; $9q^8$; $12q^2$; $20q^3$; 10; $\dfrac{-1}{q^2}$; $\dfrac{-1}{2q^{3/2}}$)

Differential coefficient of a constant term

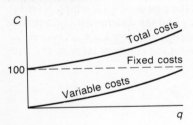

Fig. 13.3

In the examples so far considered C has been a function of q with no fixed costs. In practice there are fixed as well as variable costs, so that we can have functions such as $C = 100 + q^2$, where 100 is the fixed cost. The cost function is, therefore, made up of two elements, a fixed cost of 100 and a variable cost of q^2. These two elements and the total cost C are plotted on the same graph in Fig. 13.3. The fixed-cost line is horizontal and has, therefore, a zero gradient. The differential coefficient (i.e. rate of change) of a **constant** is **zero**.

The differential coefficient of total cost $C = 100 + q^2$ and of variable costs $C = q^2$ are both equal to $2q$. This means that for a given value of q, the gradient of the tangent to the total cost curve is the same as the gradient of the tangent to the variable cost curve.

Differentiation of a sum

If $c = q^2 + q^3 + 10$, we differentiate each term in turn, i.e.

$$\frac{dC}{dq} = 2q + 3q^2$$

Differentiation of a difference

If $C = q^3 - q^2 + 5$, then

$$\frac{dC}{dq} = 3q^2 - 2q$$

Exercise 2

Find $\dfrac{dC}{dq}$ if:

(i) $C = q^3 + q^4$; (ii) $C = q^5 - q - 4$;

(iii) $C = q^2 - \dfrac{1}{q}$; (iv) $C = 3q^3 + \dfrac{q^2}{2} - 4q$.

(*Answers:* $3q^2 + 4q^3$; $5q^4 - 1$; $2q + \dfrac{1}{q^2}$; $9q^2 + q - 4$)

Worked Example 1	If $C = 2q^2 + 4q + 10$ find an expression for marginal cost and hence find marginal cost when $q = 5$.

Marginal cost is given by

$$\frac{dC}{dq}, \text{ and } \frac{dC}{dq} = 4q + 4,$$

thus marginal cost is equal to $4q + 4$. When $q = 5$ marginal cost is equal to $4 \times 5 + 4 = 24$.

Exercise 3	If $C = q^3 - 2q^2 + 8q + 50$ find an expression for marginal cost and hence find marginal cost when $q = 7$. (*Answers:* $3q^2 - 4q + 8$; 127)

MARGINAL REVENUE

Marginal revenue may be defined as the increase in total revenue for a unit increase in sales. In calculus terms, if R is total revenue, then

marginal revenue is $\dfrac{dR}{dq}$, i.e. the rate of change of total revenue.

Worked Example 2	If the relationship between price p and quantity demanded q is $p = 14 - q$, find an expression for revenue in terms of q. Obtain an expression for marginal revenue, and find the marginal revenue when $q = 4$.

Revenue = price × quantity = $p \times q$;
but $p = 14 - q$
Revenue $(R) = (14 - q) \times q = 14q - q^2$
Marginal revenue is given by

$$\frac{dR}{dq} = 14 - 2q,$$

thus marginal revenue is given by $14 - 2q$; when $q = 4$ marginal revenue is $14 - 2 \times 4 = 6$.

Exercise 4	If $p = 30 - q$ find an expression for revenue and hence find marginal revenue; find the marginal revenue when $q = 10$. (*Answers:* $30q - q^2$; $30 - 2q$; 10)

SECOND DIFFERENTIAL COEFFICIENT

If we are given the cost function $C = q^3 - 4q^2 + 3q + 200$, we can differentiate and find

$$\frac{dC}{dq} = 3q^2 - 8q + 3.$$

It is possible to differentiate a second time; the *second differential* is written $\dfrac{d^2C}{dq^2}$ and is pronounced 'd 2 C d q squared'. To find $\dfrac{d^2C}{dq^2}$ we differentiate each term again, i.e.

$$\frac{d^2C}{dq^2} = 6q - 8.$$

Exercise 5

Find the second differential coefficient for the expressions in Exercises 1 and 2.

(*Answers:* $30q^4$; $72q^7$; $24q$; $60q^2$; 0; $\dfrac{2}{q^3}$; $\dfrac{3}{4q^{5/2}}$;

$6q + 12q^2$; $20q^3$; $2 - \dfrac{2}{q^3}$; $18q + 1$)

MAXIMA AND MINIMA

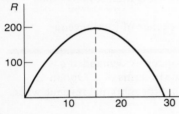

Fig. 13.4

In economics and business we are interested in maximizing revenue or profits and in minimizing costs. We are, therefore, interested in finding a mathematical technique which will enable us to do this.

Suppose we have the revenue function $R = 28q - q^2$. Which level of output would enable us to maximize revenue? We could do this graphically.

We could plot the graph as in Fig. 13.4 and note that the maximum revenue occurs when $q = 14$, giving a maximum revenue of 196. However, we can also obtain this result using calculus, since we note that at a maximum or a minimum the gradient of the tangent to the curve is zero (since the tangent is horizontal).

The gradient of the tangent to the revenue curve is given by the differential coefficient $\dfrac{dR}{dq}$; at a maximum the gradient is zero and hence

$$\frac{dR}{dq} = 0.$$

USE OF SECOND DIFFERENTIAL COEFFICIENT

In the example above, Fig. 13.4 showed that we were looking for a maximum. In examination questions you have to *show* that you have found a maximum. You could sketch the curve and this would be sufficient for most examiners, but sketching the curve is time consuming. The second differential coefficient may be used to determine whether there is a maximum or minimum. You need to **learn** the following:

If the second differential coefficient is *negative*, we have a *maximum*. If the second differential coefficient is *positive*, we have a *minimum*.

Worked Example 3	If the revenue function is $R = 28q - q^2$ find the level of output which maximizes revenue; find also the maximum revenue.

$$R = 28q - q^2$$

$$\frac{dR}{dq} = 28 - 2q \qquad \frac{d^2R}{dq^2} = -2$$

As $\dfrac{d^2R}{dq^2} = -2$ is negative, we have a maximum.

At a maximum $\dfrac{dR}{dq} = 0 \therefore 0 = 28 - 2q$

$$\therefore 2q = 28 \therefore q = 14.$$
The maximum revenue is $R = 28 \times 14 - 14^2 = 196$.

Exercise 6

For the cost function $C = 64 - 8q + q^2$ find the level of output which minimizes costs and find the minimum costs.
(*Answers:* 4; 48)

Worked Example 4

Given the revenue function $R = 28q - q^2$ and the cost function $C = 64 - 8q + q^2$, find an expression for the profit function and hence find the output which maximizes profits. Find also the corresponding price at profit maximization.

$$\begin{aligned}
&\text{Profits} = \text{revenue } (R) - \text{costs } (C) \\
&P = 28q - q^2 - (64 - 8q + q^2) \\
&P = -64 + 36q - 2q^2
\end{aligned}$$

$$\frac{dP}{dq} = 36 - 4q \qquad \frac{d^2P}{dq^2} = -4.$$

As $\dfrac{d^2P}{dq^2} = -4$, this is negative; therefore a maximum.

At a maximum $\dfrac{dP}{dq} = 0 \therefore 0 = 36 - 4q \therefore q = 9$.

Maximum profits $= -64 + 36 \times 9 - 2 \times 9^2 = 98$.
To obtain the *price* we note that revenue is equal to price × quantity, i.e. $R = p \times q$.
Now $R = 28q - q^2$, thus $28q - q^2 = pq$.
Dividing both sides by q we obtain

$$p = 28 - q.$$

At profit maximization $q = 9$;

thus $p = 28 - 9 = 19$.

If the cost and revenue functions are $R = 40q - q^2$ and

$$C = 20 + 4q + \frac{q^2}{2},$$

find an expression for profit and hence find the output which maximizes profit, the maximum profit and the price at profit maximization.

(*Answers:* $-20 + 36q - \frac{3q^2}{2}$; 12; 196; 28)

MARGINAL REVENUE AND MARGINAL COST

A well-known result in economics is that *profit maximization* occurs when *marginal revenue equals marginal cost*. Consider the revenue and cost functions given in Worked Example 4.

$$R = 28q - q^2 \qquad C = 64 - 8q + q^2$$

$$\frac{\mathrm{d}R}{\mathrm{d}q} = 28 - 2q \qquad \frac{\mathrm{d}C}{\mathrm{d}q} = -8 + 2q$$

At profit maximization, marginal revenue = marginal cost

$$\therefore\ 28 - 2q = -8 + 2q$$
$$\therefore\ 36 = 4q$$
$$\therefore\ q = 9, \text{ as found in Worked Example 4.}$$

CURVES WITH BOTH A MAXIMUM AND A MINIMUM

Worked Example 5

If we have the cost curve

$$C = \frac{q^3}{3} - 4q^2 + 12q + 150,$$

find the values of q which give a maximum and a minimum and find the corresponding values of C.

If $C = \frac{q^3}{3} - 4q^2 + 12q + 150,$

then $\dfrac{\mathrm{d}C}{\mathrm{d}q} = q^2 - 8q + 12$

and $\dfrac{\mathrm{d}^2C}{\mathrm{d}q^2} = 2q - 8.$

At a maximum or minimum, $\dfrac{\mathrm{d}C}{\mathrm{d}q} = 0$

$$q^2 - 8q + 12 = 0 \ \therefore\ (q - 6)(q - 2) = 0 \ \therefore\ q = 6 \text{ or } q = 2$$

Maximum or minimum occur at 6 and 2.

When $q = 2$, $\frac{d^2C}{dq^2} = 2 \times 2 - 8 = -4$, which is negative.

Hence a maximum and $C = 160.67$ on substituting $q = 2$ in cost equation.

When $q = 6$, $\frac{d^2C}{dq^2} = 2 \times 6 - 8 = +4$, which is positive.

Hence a minimum and $C = 150$ on substituting $q = 6$ in cost equation.

Exercise 8

The total cost curve of a firm in terms of the output q is given by the expression $C = q^3 - 6q^2 + 9q + 120$. Find the levels of output which give a maximum and minimum, and the corresponding costs.
(*Answers:* minimum 3, cost 120; maximum 1, cost 124)

Notes
(i) It is essential to remember that *for a maximum the second differential coefficient is negative; for a minimum the second differential coefficient is positive*.
(ii) To be sure of full marks on maxima and minima questions it is essential to find a second differential coefficient and to establish that you have found a maximum or a minimum.

ELASTICITY OF DEMAND

Price elasticity of demand is defined as the ratio

$$E = \frac{\text{Proportionate change in quantity demanded}}{\text{Proportionate change in price of commodity}}$$

$$E = \frac{\frac{\Delta q}{q}}{\frac{\Delta p}{p}} = \frac{\Delta q}{q} \times \frac{p}{\Delta p} = \frac{p}{q} \times \frac{\Delta q}{\Delta p}$$

or in calculus terms $E = \frac{p}{q} \times \frac{dq}{dp}$

Worked Example 6

If $q = 10 - 2p$, find an expression for price elasticity of demand and evaluate this when $p = 3$.

$$E = \frac{p}{q} \times \frac{dq}{dp}, \qquad q = 10 - 2p, \text{ thus } \frac{dq}{dp} = -2.$$

Hence $E = \frac{p}{q} \times -2 = \frac{-2p}{q}$.

It is usual to take the numerical value $E = \frac{2p}{q}$.

When $p = 3$, $q = 10 - 2p = 10 - 2 \times 3 = 4$.

Thus $E = \dfrac{2 \times 3}{4} = \dfrac{3}{2}$.

Exercise 9

(a) Obtain an expression for price elasticity of demand in terms of q for each of the following demand functions.

(i) $q = 20 - 4p$; (ii) $q = 10 - \dfrac{p}{2}$; (iii) $q = \dfrac{5}{p}$.

(b) For each expression in part (a) find the quantity demanded when the elasticity of demand equals -1.

(*Answers:* $\dfrac{q-20}{q}$; $\dfrac{q-10}{q}$; -1; 10, 5, any quantity)

OTHER CALCULUS TOPICS

Note: This chapter has now covered the usual topics of calculus set in most examinations; the topics of partial differentiation and integration in the following sections can be ignored by most examinees. However, these topics have been covered in examination questions set by the CACA (ACCA) and again feature in undergraduate and some diploma courses.

PARTIAL DIFFERENTIATION

In this chapter we have assumed that costs were a function of output, i.e. of one variable q. In some situations a function may depend on *more than one* variable. This is the usual case with production, where the volume of production P depends on the labour input L and the capital input K.

Thus $P = f(L,K)$.

This is called a *production function*. An example of a production function is

$$P = 16LK - 4L^2 - 2K$$

The *partial differential coefficient* $\dfrac{\partial P}{\partial L}$ shows the change in the volume of production resulting from a unit change in the labour force L with constant capital. Thus when we differentiate *partially* with respect to L, we assume K is a constant. We follow the *general rule* for differentiation, but remembering that we treat K as a constant.

Hence $\dfrac{\partial P}{\partial L} = 16K - 8L$.

The last term disappears as $2K$ is a constant.

Similarly, $\dfrac{\partial P}{\partial K}$ shows the change in the volume of production resulting from a unit change in K with L constant. When we differentiate *partially* with respect to K, we assume L is a constant.

Hence $\dfrac{\partial P}{\partial L} = 16L - 2$.

The middle term disappears as L is a constant.

Worked Example 7

If $P = 12LK - L^2 - 3K^2$ find $\dfrac{\partial P}{\partial L}$ and $\dfrac{\partial P}{\partial K}$.

$$\dfrac{\partial P}{\partial L} = 12K - 2L \qquad \dfrac{\partial P}{\partial K} = 12L - 6K$$

Exercise 10

If $P = 8LK - 2L^2 - 3K^2 + 2L + 8$ find $\dfrac{\partial P}{\partial L}$ and $\dfrac{\partial P}{\partial K}$.

(*Answers:* $8K - 4L + 2$; $8L - 6K$)

INTEGRATION

Integration may be considered as the reverse of differentiation.

If $C = 100 + q^2$ then $\dfrac{dC}{dq} = 2q$.

If $\dfrac{dC}{dq} = 2q$ what is the value of C?

By reversing the differentiation process we have $C = q^2$. However, this is not the same as $C = 100 + q^2$ since the fixed costs of 100 are omitted. When we differentiate a constant term we always obtain zero, so when we *reverse* the process of differentiation, i.e. when we *integrate*, we always have to add a constant.

Thus if $\dfrac{dC}{dq} = 2q$ then $C = q^2 + K$, where K is a constant.

When we integrate we use the integral sign $\int$.
If we wish to integrate $2q$ we write $\int 2q \, dq$:

$$\int 2q \, dq = q^2 + K.$$

In some situations it is possible to evaluate the constant K.

Worked Example 8

A company finds that the marginal revenue is given by the expression $20 - 2q$; obtain the revenue function. The marginal cost is given by the expression $4q - 10$; if the fixed costs are 30, obtain the cost function.

Marginal revenue is $\dfrac{dR}{dq}$, thus $\dfrac{dR}{dq} = 20 - 2q$.

$$R = \int (20 - 2q) \, dq$$
$$R = 20q - q^2 + K$$

When $q = 0$, i.e. when there is nothing to sell, there is no revenue, thus R must be zero. If we substitute $q = 0$ and $R = 0$ we find $K = 0$. Thus the revenue function is:

$$R = 20q - q^2$$

Marginal cost is $\dfrac{dC}{dq}$, thus $\dfrac{dC}{dq} = 4q - 10$

$$C = \int (4q - 10)\,dq$$

$$C = 2q^2 - 10q + K$$

When $q = 0$ there are no variable costs, but the fixed costs are 30, thus when $q = 0$, $C = 30$. Substituting, $30 = 0 - 0 + K \therefore K = 30$. Thus the cost function is:

$$C = 2q^2 - 10q + 30$$

Exercise 11

(i) If the marginal costs are $2q - 20$, and fixed costs are 15, find the cost function.

(ii) If the marginal revenue is given by $20 - 4q$, find an expression for the revenue function.

(*Answers:* $C = q^2 - 20q + 15$; $R = 20q - 2q^2$)

C. SOLUTIONS TO EXERCISES

S1

We use $\dfrac{dC}{dq} = nq^{n-1}$.

(i) $n = 6, \dfrac{dC}{dq} = 6q^{6-1} = 6q^5$;

(ii) $n = 9, \dfrac{dC}{dq} = 9q^{9-1} = 9q^8$;

(iii) $\dfrac{dC}{dq} = 4 \times 3q^{3-1} = 12q^2$;

(iv) $\dfrac{dC}{dq} = 5 \times 4q^{4-1} = 20q^3$;

(v) $\dfrac{dC}{dq} = 10 \times 1q^{1-1} = 10 \ (q^0 = 1)$;

(vi) $n = -1, \dfrac{dC}{dq} = -1q^{-1-1} = -1q^{-2} = \dfrac{-1}{q^2}$;

(vii) $n = -\dfrac{1}{2}, \dfrac{dC}{dq} = -\dfrac{1}{2} q^{-\frac{1}{2}-1} = -\dfrac{1}{2q^{3/2}}$.

S2

(i) $\dfrac{dC}{dq} = 3q^{3-1} + 4q^{4-1} = 3q^2 + 4q^3$;

(ii) $\dfrac{dC}{dq} = 5q^{5-1} - 1q^{1-1} = 5q^4 - 1$;

(iii) $C = q^2 - 1q^{-1}$, $\dfrac{dC}{dq} = 2q^{2-1} - (-1q^{-1-1}) = 2q + \dfrac{1}{q^2}$;

(iv) $\dfrac{dC}{dq} = 3 \times 3q^{3-2} + \frac{1}{2} \times 2q^{2-1} - 4 = 9q^2 + q - 4$.

S3

$\dfrac{dC}{dq} = 3q^2 - 4q + 8$; put $q = 7$: $3 \times 49 - 4 \times 7 + 8 = 127$

S4

$R = pq = (30 - q)q = 30q - q^2$; $\dfrac{dR}{dq} = 30 - 2q$; when $q = 10$, $\dfrac{dR}{dq} = 10$

S5

$\dfrac{d^2C}{dq^2} = 6 \times 5q^{5-1} = 30q^4$; $\dfrac{d^2C}{dq^2} = 9 \times 8q^{8-1} = 72q^7$;

$\dfrac{d^2C}{dq^2} = 12 \times 2q = 24q$; $\dfrac{d^2C}{dq^2} = 20 \times 3q^{3-2} = 60q^2$; $\dfrac{d^2C}{dq^2} = 0$;

$\dfrac{d^2C}{dq^2} = -1 \times -2q^{-2-1} = \dfrac{2}{q^3}$; $\dfrac{d^2C}{dq^2} = \dfrac{-1}{2} \times \dfrac{-3}{2} q^{\frac{-3}{2}-1} = \dfrac{3}{4} q^{5/2}$;

$\dfrac{d^2C}{dq^2} = 3 \times 2q^{2-1} + 4 \times 3q^{3-1} = 6q + 12q^2$; $\dfrac{d^2C}{dq^2} = 5 \times 4q^{4-1} + 0 = 20q^3$;

$\dfrac{d^2C}{dq^2} = 2 + (-2)q^{-2-1} = 2 - \dfrac{2}{q^3}$; $\dfrac{d^2C}{dq^2} = 9 \times 2q^{2-1} + 1 = 18q + 1$

S6

$\dfrac{dC}{dq} = -8 + 2q$, $\dfrac{d^2C}{dq^2} = +2$, positive, thus minimum.

$\dfrac{dC}{dq} = 0 = -8 + 2q$, $q = 4$; $C = 64 - 8 \times 4 + 4^2 = 48$.

S7

$P = R - C = 40q - q^2 - \left(20 + 4q + \dfrac{q^2}{2}\right) = -20 + 36q - \dfrac{3q^2}{2}$

$\dfrac{dP}{dq} = 36 - 3q$, $\dfrac{d^2P}{dq^2} = -3$, negative, thus maximum. $0 = 36 - 3q$, $q = 12$.

$P = 196$; $pq = 40q - q^2$, $p = 40 - q$, $p = 40 - 12 = 28$.

S8

$$\frac{dC}{dq} = 3q^2 - 12q + 9, \frac{d^2C}{dq^2} = 6q - 12$$

$$0 = 3q^2 - 12q + 9, \ 0 = q^2 - 4q + 3, \ 0 = (q-3)(q-1), \ q = 3, \ q = 1$$

$$q = 3, \frac{d^2C}{dq^2} = 6 \times 3 - 12 = +6, \text{ positive, minimum; put } q = 3, \ C = 120$$

$$q = 1, \frac{d^2C}{dq^2} = 6 \times 1 - 12 = -6, \text{ negative, maximum; put } q = 1, \ C = 124$$

S9

(a) (i) $\dfrac{dq}{dp} = -4, \ E = \dfrac{p}{q} \times -4, \ q = 20 - 4p, \ -4p = q - 20, \ E = \dfrac{q-20}{q}$

(ii) $\dfrac{dq}{dp} = \dfrac{-1}{2}, \ E = \dfrac{p}{q} \times \dfrac{-1}{2}, \ q = 10 - \dfrac{p}{2}, \ \dfrac{-p}{2} = q - 10, \ E = \dfrac{q-10}{q}$

(iii) $\dfrac{dq}{dp} = \dfrac{-5}{p^2}, \ E = \dfrac{p}{q} \times \dfrac{-5}{p^2} = \dfrac{-5}{qp}, \ q = \dfrac{5}{p}, \ qp = 5, \ E = \dfrac{-5}{5} = -1$

(b) When $E = -1$, (i) $-1 = \dfrac{q-20}{q}, \ q = 10$; (ii) $-1 = \dfrac{q-10}{q}, \ q = 5$;

(iii) For this expression elasticity is -1 for all quantities.

S10

$$\frac{\partial p}{\partial L} = 1 \times 8L^{1-1}K - 2 \times 2L^{2-1} + 1 \times 2L^{1-1},$$

using our general rule for differentiating p with respect to L, treating K as a constant, i.e.

$$\frac{\partial p}{\partial L} = 8K - 4L + 2$$

Similarly, $\dfrac{\partial p}{\partial k} = 1 \times 8LK^{1-1} - 2 \times 3K^{2-1} = 8L - 6K$

S11

(i) $\dfrac{dC}{dq} = 2q - 20, \ C = \int(2q - 20)dq, \ C = q^2 - 20q + K.$

When $q = 0 \ C = 15$, thus $K = 15$; $C = q^2 - 20q + 15$

(ii) $\dfrac{dR}{dq} = 20 - 4q, \ R = \int(20 - 4q)dq, \ R = 20q - 2q^2 + K.$

When $q = 0$ then $R = 0$, hence $K = 0$; $R = 20q - 2q^2$

D. RECENT EXAMINATION QUESTIONS

Q1

The demand for a product is given by the following equation:

$$p = 30 - q$$

where p is the price and q is the quantity sold.

The relationship between costs (C) and output q is given by:

$$C = 4q^2 - 80q + 450$$

Use the above expressions to find:
(i)   the revenue function,
(ii) the profit function,
(iii) maximum revenue,
(iv) minimum costs,
(v)   maximum profits.

(*Answers:* $30q - q^2$; $110q - 5q^2 - 450$; 225; 50; 155)

Q2

A demand function is given by $p = x^2 - 24x + 117$, where x units is the quantity demanded and £p the price per unit.
(a) Write down an expression for the total revenue for x units of production.
(b) Using the methods of differential calculus establish the number of units of production and the price at which total revenue will be maximized.
(c) If elasticity of demand is defined as $(p/x)\,(1/(\mathrm{d}p/\mathrm{d}x))$ determine the elasticity of demand for the quantity which maximizes the total revenue.

(CACA (ACCA) (part) June 1984)

(*Answers:* $x^3 - 24x^2 + 117x$; 3, 54; -1)

Q3

(a) Demand is given by $q = 2p^3 - 21p^2 + 36p + 9$.
Using the method of differential calculus establish the maximum demand and its corresponding price.
(b) Elasticity of demand is given by the expression

$$\frac{\mathrm{d}q}{\mathrm{d}p}\,\frac{p}{q}.$$

If the demand function is given by

$$q = \frac{90}{p^3},$$

show that the elasticity of demand is constant.

(CACA (ACCA) (part) Dec. 1984)

(*Answers:* 26, 1; -3)

Q4

Your company manufactures large-scale items. It has been shown that the marginal cost, which is the gradient of the total cost curve, is $(92-2x)£$ thousands, where x is the number of units of output per annum. The fixed costs are £800,000 per annum. It has also been shown that the marginal revenue, which is the gradient of the total revenue curve, is $(112-2x)£$ thousands.

(a) Establish by integration the equation of the total cost curve.
(b) Establish by integration the equation of the total revenue curve.
(c) Establish the break-even situation for your company.
(d) Determine the number of units of output that would:
    (i)   maximize the total revenue, and
    (ii) *maximize* the total costs, together with the maximum total revenue and total costs.
(e) Assuming that your company cannot manufacture more than 60 units of output per annum, what interpretation can be put on the results you obtain in (d)? (A sketch of the total revenue and cost curve will be helpful.)

(CACA (ACCA) Dec. 1982)

(*Answers:* $800+92x-x^2$; $112x-x^2$; 40; 56, £3,316,000; 46, £2,916,000)

E. OUTLINE ANSWERS TO EXAM QUESTIONS

A1

(i) $R=p\times q=(30-q)q=30q-q^2$;

(ii) $P=R-C, P=30q-q^2-(4q^2-80q+450)=110q-5q^2-450$;

(iii) $\dfrac{dR}{dq}=30-2q, \dfrac{d^2R}{dq^2}=-2$, negative, thus max. $0=30-2q$, $q=15$,

       $R=30\times15-15^2=225$;

(iv) $\dfrac{dC}{dq}=8q-80, \dfrac{d^2C}{dq^2}=8$, positive, thus minimum.

       $0=8q-80$, $q=10$, $C=4\times10^2-80\times10+450=50$;

(v) $\dfrac{dP}{dq}=110-10q, \dfrac{d^2P}{dq^2}=-10$, negative, thus max.

      $0=110-10q$, $q=11$, $P=110\times11-5\times11^2-450=155$

A2

(a) $R=x(x^2-24x+117)=x^3-24x+117x$

(b) $\dfrac{dR}{dx}=3x^2-48x+117, \dfrac{d^2R}{dx^2}=6x-48$;

      $0=3x^2-48x+117, 0=x^2-16x+39, 0=(x-3)(x-13)$,
      $x=3, x=13$.

When $x=3$, $\frac{d^2R}{dx^2}=6\times 3-48=-30$, negative, thus max.

$p=3^2-24\times 3+117=54$

(c) $p=x^2-24x+117$, $\frac{dp}{dx}=2x-24$.

When $x=3$, $\frac{dp}{dx}=2\times 3-24=-18$.

$E=(p/x)\,(1/(dp/dx))=\frac{54}{3}\times\frac{1}{-18}=-1$

A3

(a) $\frac{dq}{dp}=6p^2-42p+36$, $\frac{d^2q}{dp^2}=12p-42$.

$0=6p^2-42p+36$, $0=p^2-7p+6$, $0=(p-6)(p-1)$;

$p=6$, $p=1$, when $p=1$, $\frac{d^2q}{dp^2}=-30$, negative, thus max.

$q=2-21+36+9=26$

(b) $\frac{dq}{dp}=\frac{-3\times 90}{p^4}=\frac{-270}{p^4}$

$\frac{dq}{dp}\times\frac{p}{q}=\frac{-270}{p^4}\times\frac{p}{q}=\frac{-270}{p^4}\times\frac{p}{90/p^3}=\frac{-270}{90}=-3$, which is a constant.

A4

(a) $\frac{dC}{dx}=92-2x$, $C=\int(92-2x)dx$, $C=92x-x^2+K$.

When $x=0$, $C=800$ (units are £1,000); thus $K=800$, $C=92x-x^2+800$

(b) $\frac{dR}{dx}=112-2x$, $R=\int(112-2x)dx$, $R=112x-x^2+K$.

When $x=0$, $R=0$; thus $K=0$, $R=112x-x^2$

(c) At break-even $R=C$; $92x-x^2+800=112x-x^2$, $20x=800$, $x=40$

(d) $R=112x-x^2$, $\frac{dR}{dx}=112-2x$, $\frac{d^2R}{dx^2}=-2$,

negative, thus maximum. $0=112-2x$, $x=56$

$R=112\times 56-56^2=3,316$. Thus maximum revenue is £3,316,000.

$\frac{dC}{dx}=92-2x$, $\frac{d^2C}{dx^2}=-2$, negative, thus max. $0=92-2x$, $x=46$,

$C = 92 \times 46 - 46^2 + 800 \doteq 2{,}916$. Thus maximum costs are £2,916,000.

(e) Total costs start to fall at 46; revenue is maximized at 56. At 60 revenue curve has a gradient of $112 - 2 \times 60 = -8$; cost curve has a gradient $92 - 2 \times 60 = -28$. Hence costs are reducing faster than revenue so that maximum profit occurs at output of 60. (See Fig. 13S.1.)

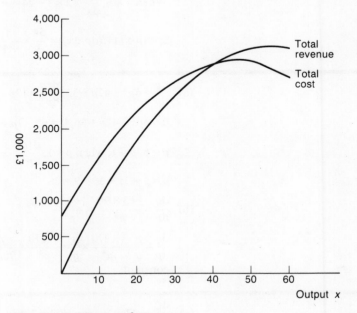

Fig. 13S.1   Costs and revenue

A STEP FURTHER   Tennant-Smith, *Mathematics for the Manager*, Chs 9, 10 and 11.

Chapter 14

Financial mathematics

A. GETTING STARTED

It is essential to check whether financial tables are provided in the examination room – if they are try to obtain a copy. Calculators are essential for this topic. Although financial calculators can be purchased which considerably simplify the work, these calculators may not be allowed. Most examinations require you to *show* the necessary steps in the calculation, so that a sophisticated calculator which just gives the final result would be of limited use.

B. ESSENTIAL PRINCIPLES

INTEREST

In business, investment decisions have to be made. Such decisions usually involve an initial capital payment, with the hope that a flow of profits will follow in the future. In making decisions which involve future payments and/or receipts it is necessary to consider the *rate of interest*.

The rate of interest is denoted by i and is defined as follows:

$$i = \frac{\text{interest payable for the period concerned}}{\text{capital at the beginning of the period}}$$

If interest of £25 is payable each year in respect of a loan of £500, the annual rate of interest is

$$i = \frac{25}{500} = 0.05.$$

Interest is often quoted as a rate per 100 units of capital, i.e. as a rate per cent. In this example the rate per cent is $100 \times 0.05 = 5\%$.

In all compound interest calculations the decimal form *i* should be used. If you are given a *percentage* rate of interest you must convert it to the decimal form by dividing by 100.

If P_0 is the initial principal, i.e. the initial capital, and i is the rate of interest, then the interest payable is $P_0 \times i$.

Worked Example 1

Find the yearly interest on a loan of £750 at a rate of interest of 8%.

Here $P_0 = 750$ and $i = 8/100 = 0.08$. Yearly interest is $P_0 \times i = 750 \times 0.08 = £60$.

Exercise 1

A man invested £700 and received £42 interest payment at the end of the year. Calculate the rate of interest (i) as a decimal, (ii) as a percentage. If he had invested £1,200 at the same rate of interest, what interest payment would he receive at the end of the year?
(*Answers:* 0.06; 6%; £72)

SIMPLE AND COMPOUND INTEREST

Suppose £10,000 is lent to Mr A at 10% p.a. for 3 years. At the end of each year £1,000 interest is payable and the £10,000 is returned at the end of the third year. The investor could take his £1,000 each year, or he could have agreed with Mr A to collect the total interest of £3,000 at the end of the 3 years. Such an agreement is called a **simple** interest transaction.

Capital and accrued interest	£10,000	£11,000	£12,000	£13,000
Year	0	1	2	3
Interest		£1,000	£1,000	£1,000

Alternatively, the investor could have agreed with Mr A to have the interest at the end of each year added to the principal and in subsequent years to receive interest on the principal and the accrued interest. Thus at the beginning of year 2 the principal plus interest is £10,000 + £1,000 = £11,000. In year 2 the investor receives £11,000 × 0.1 = £1,100 interest. At the beginning of year 3 the principal plus interest is £11,000 + £1,100 = £12,100. In year 3 the investor receives £12,100 × 0.1 = £1,210 interest. At the end of year 3 the principal plus interest is £12,100 + £1,210 = £13,310. Here interest has been compounded into principal. Such an arrangement is called a **compound** interest transaction.

Capital and accrued interest	£10,000	£11,000	£12,100	£13,310
Year	0	1	2	3
Interest		£1,000	£1,100	£1,210

In practice, simple interest transactions seldom occur. In examination questions *always assume compound interest* unless you are *specifically* told that the transaction is *simple interest*.

There is a formula which allows the principal to be found for a given rate of interest. If P_0 is the initial principal, P_n is the principal at the end of *n* years and *i* is the rate of interest, then

$$P_n = P_0(1 + i)^n$$

Special tables are available which work out $(1 + i)^n$ for certain values of i and n. A limited set of financial tables are to be found on p. 245. If tables are not supplied or if the tables do not have the value(s) of i and n needed, then calculators have to be used. Suppose that you want to find the value of $(1 + i)^n$ when the rate of interest is 10% and the number of years is 6; here $i = 0.1$ and $n = 6$, and we need to find $(1.1)^6$. If your calculator has an x^y key, the usual procedure is to enter 1.1, press the x^y key, then enter 6 and then press the $=$ key: you should obtain 1.771561. If you have a calculator without an x^y key, the following procedure usually works; enter 1.1, press the $\times$ key twice (for some calculators once), then the $=$ key, this gives 1.21 $(= 1.1^2)$. If you press the $=$ key again you obtain $1.1^3 = 1.331$, and on pressing the $=$ key yet again you obtain $1.1^4 = 1.4641$; continuing this procedure you will find that $1.1^6 = 1.771561$. If you look at Table 1 on p. 245 you will find that for compound interest at 10% over 6 years the value is 1.77156.

| **Worked Example 2** | £7,000 is invested at 15%. Find the amount accumulated at the end of (i) 5 years; (ii) 10 years. |

(i) We can use the financial tables on p. 245, for 15% and 5 years we obtain 2.01136. The amount is therefore $7,000 \times 2.01136$ $= £14,079.52$.

(ii) Here $n = 10$ and $i = 15/100 = 0.15$; thus $1 + i = 1 + 0.15 = 1.15$. Since $1.15^{10} = 4.0455577$, the amount is $7,000 \times 4.0455577 = £28,318.90$.

| **Exercise 2** | £6,000 is invested for (a) 4 years; (b) 12 years. If the rate of interest is: (i) 5%; (ii) 10%; (iii) 7.5%, find the amount. *(Answers:* 7,293.06; 8,784.60; 8,012.81; 10,775.14; 18,830.57; 14,290.68) |

ANNUALIZED PERCENTAGE RATE

Recent legislation states that when a rate of interest is quoted in respect of a credit agreement, the interest must be given at an *annualized percentage rate* (APR). In the case of some credit cards the interest is compounded monthly, and the rate may, for example, be quoted as 1.75% per month. Nevertheless the credit card company must also state the APR.

If i is the *monthly* rate of interest, then £1 would increase to £1 $\times (1 + i)^{12}$ at the end of the year. Thus the interest at an *annual* rate is $(1 + i)^{12} - 1$. To obtain the APR multiply by 100. Similarly, if interest is compounded *half-yearly* (most building societies do this) then APR is $100 \times ((1 + i)^2 - 1)$.

| **Worked Example 3** | A credit card company charges 2% per month, compounded monthly. Calculate the APR. |

$$i = 2/100 = 0.02. \text{ Thus } (1 + i)^{12} = 1.02^{12} = 1.2682418$$

Hence APR is $100 \times (1.2682418 - 1) = 26.82\%$.

A man requires a loan of £1,000 to be repaid in full at the end of the year. A finance company offers two methods of charging interest: (i) compounded monthly at a rate of 2.25% per month; (ii) compounded six-monthly at a rate of 14% per six months. Calculate the APR in both cases.
(*Answers:* 30.61; 29.96)

SINKING FUND

Suppose that at the end of each year a businessman placed an amount £5,000 into a special fund (this is usually called a *sinking fund*). We need to find out the amount of the fund at the end of (say) 4 years assuming that the rate of interest is a constant 10%.

At the end of the first year the fund receives £5,000. At the end of the second year interest of £5,000 × 0.1 = £500 is added, together with a further £5,000, making the value of the fund £10,500. At the end of the third year interest of £10,500 × 0.1 = £1,050 is added, together with a further £5,000, making the total £16,550. At the end of the fourth year interest of £16,550 × 0.1 = £1,655 is added, together with a further £5,000, making the total £23.205.

It is possible to use Table 3 on page 245 to find this result. We obtain $s_{\overline{n}|}$ for 10% and year 4 = 4.641 and multiply by 5,000, i.e. 4.641 × 5,000 = £23,205.

There is a formula for $S_{\overline{n}|}$ which is obtained by summing the geometric progression $A + A(1 + i) + A(1 + i)^2 + \ldots + A(1 + i)^{n-1}$:

$$S_{\overline{n}|} = \frac{A((1 + i)^n - 1)}{i}$$

where A is the annual payment, i is rate of interest, and n the number of years.

Examination candidates tend to apply the formula or use the tables without thought. The tables and the formula assume payments into the sinking fund are made at the **end** of the year. If the examiner sets the question with payments at the **beginning** of the year, far too many candidates obtain the wrong result. If payments are made at the **beginning** of the year then each payment earns an **extra** year's interest. In the example above the £23,205 is increased by 10% (the rate of interest), i.e. the amount is 23,205 × 1.1 = £25,525.50.

Worked Example 4

At the end of each year a firm places £2,000 in a sinking fund. Find the amount in the sinking fund at the end of 6 years if the rate of interest is 15%.

We can use the tables: $S_{\overline{6}|} = 8.75374$; 2,000 × 8.75374 = £17,507.

Worked Example 5

A man invests at the beginning of each year £200 in a building society. If the rate of interest is 7.8%, find the amount at the end of 5 years.

As the investment is at the **beginning** of each year the formula is

$$(1+i)AS_{\overline{n}|} \text{ or } (1+i)\frac{A((1+i)^n-1)}{i}$$

However, as 7.8% is not shown in the tables we have to use the formula. $A = 200$, $n = 5$, $i = 7.8/100 = 0.078$, $1 + i = 1.078$.

$$1.078 \times 200 \times \frac{(1.078^5 - 1)}{0.078} = 215.6 \times \frac{(1.455773 - 1)}{0.078} = £1,259.80$$

Exercise 4

£1,000 is invested at the end of each year for 6 years. Find the amount at the end of 6 years if the rate of interest is: (i) 10%; (ii) 6%. (*Answers:* £7,715.61; £6,975.32)

Exercise 5

£1,000 is invested at the beginning of a year for each of 5 years. Find the amount accumulated if the rate of interest is (i) 10%; (ii) 7%. (*Answers:* £6,715.61; £6,153.29)

PRESENT VALUE

Suppose that a firm wishes to replace some equipment in 5 years' time. It expects that the equipment will cost £10,000. How much would the firm have to invest today so that it will have £10,000 in 5 years' time if the rate of interest is 10%?

Let P be the amount. If i is rate of interest, then $P(1 + i)^5 = 10,000$; hence

$$P = \frac{10,000}{(1+i)^5}$$

Here $i = 0.1$, thus $1 + i = 1.1$.

Hence $P = \dfrac{10,000}{1.1^5} = \dfrac{10,000}{1.61051} = £6,209.21$.

Thus £6,209.21 invested today would yield £10,000 in 5 years time. We say that £6,209.21 is the **present value** at 10% of £10,000 due in 5 years.

In general, the present value, at a rate i, of an amount S due in n years time, is

$$\frac{S}{(1+i)^n}$$

The operation of taking the present value of a future amount is called **discounting**. The present value is sometimes called the *discounted value*; i is sometimes called the *discount rate*. The quantity $\dfrac{1}{(1+i)^n}$ is called the *discount factor*.

Table 2 on p. 245 gives the discount factors. For the example above, the table shows that for 10% and 5 years, the discount factor is 0.62092. Thus the present value of £10,000 is $10,000 \times 0.62092 = £6,209.20$.

Worked Example 6

Find the present value of £3,000 due in 6 years time at a rate of discount of 15%.

From the tables the discount factor is 0.43233, thus the amount is $3{,}000 \times 0.43233 = £1{,}297$.

Exercise 6

Find the present value of £5,500 due in 5 years at a discount rate of:
(i) 5%; (ii) 7%.
(*Answers:* £4,309; £3,921)

ANNUITIES

An *annuity* is a fixed sum paid periodically. A *fixed annuity* is an annuity paid for a fixed period: say 10 years. A *life annuity* is an annuity which is payable during the lifetime of a person. A *deferred annuity* is an annuity which does not begin until after a certain number of years. A *perpetual annuity* is an annuity which continues for ever.

Present value of an annuity

Suppose that each year an amount A is paid out and that the process is repeated for n years. Let the discount rate be i.

The payment A at the *end* of the first year has a present value of $A/(1+i)$ at the *beginning* of the first year. The payment A at the *end* of the second year has a *present value* of $A/(1+i)^2$ at the beginning of the *first* year. Continuing this process, the present value of the annuity is:

$$PV = \frac{A}{(1+i)} + \frac{A}{(1+i)^2} + \frac{A}{(1+i)^3} + \ldots + \frac{A}{(1+i)^n}$$

If A and i are known, the present value can be found. There is a formula which can be used:

$$PV = \frac{A(1-(1+i)^{-n})}{i} = Aa_{\overline{n}|}$$

Table 4 on p. 245 finds $a_{\overline{n}|}$ for various values of i and n.

Worked Example 7

Find the present value of an annuity of £1,000 p.a. payable at the end of each year when the discount rate is 10% if the annuity is for (i) 5 years; (ii) perpetual.

If the annuity were paid at the beginning of the year recalculate the result for 5 years.

$$i = 0.1; \quad (1+i) = 1.1$$

$$PV = \frac{1{,}000}{1.1} + \frac{1{,}000}{1.1^2} + \frac{1{,}000}{1.1^3} + \frac{1{,}000}{1.1^4} + \frac{1{,}000}{1.1^5}$$

$$= \frac{1{,}000}{1.1} + \frac{1{,}000}{1.21} + \frac{1{,}000}{1.331} + \frac{1{,}000}{1.4641} + \frac{1{,}000}{1.61051}$$

$$= 909.09 + 826.45 + 751.31 + 683.01 + 620.92 = £3{,}790.78.$$

If we use Table 4 on p. 245 we find that for 10% and 5 years $a_{\overline{5}|}$

$= 3.79079$; thus the present value is $1,000 \times 3.79079 = £3,790.79$ (the penny difference is due to rounding errors).

The formula for a *perpetual annuity* is

$$\frac{A}{i} = \frac{1,000}{0.1} = £10,000.$$

If the annuity were paid at the *beginning* of the year then

$$PV = 1,000 + \frac{1,000}{1.1} + \frac{1,000}{1.21} + \frac{1,000}{1.331} + \frac{1,000}{1.4641}$$

$$= 1,000 + 909.09 + 826.45 + 751.31 + 683.01 = £4,169.86$$

The value of £4,169.86 can be found by multiplying the previous result by $(1 + i)$, i.e. by 1.1; thus $£3,790.78 \times 1.1 = £4,169.86$.

Exercise 7

Find the present value of an annuity of £2,000 p.a. when the rate of discount is (i) 15%, (ii) 7%, if the annuity is payable for 6 years (a) at end of year, (b) at beginning of year.
(*Answers:* £7,568.97; £8,704.32; £9,533.08: £10,200.40)

MORTGAGES

When a house is purchased, it is usual to take out a mortgage for £P. Most *mortgages* are arranged so that the repayment of capital and interest payments are a constant annual amount £A. The present value of these future payments A must equal the mortgage P.

$$\text{Thus } P = Aa_{\overline{n}|} \text{ or } P = A\frac{(1 - (1 + i)^{-n})}{i}$$

Worked Example 8

A man takes out a mortgage for £20,000 for 25 years. If the rate of interest is 12%, what is the annual mortgage payment?

$$P = A\frac{(1 - (1 + i)^{-n})}{i}$$

The first step is to find $(1 + i)^{-25}$ which is equal to $1/(1 + i)^{25}$.
$i = 0.12$, $(1 + i) = 1.12$, $1/1.12^{25} = 1/17.000064 = 0.0588233$.
Thus $20,000 = A(1 - 0.0588233)/0.12 = A \times 7.8431391$

Hence $A = \dfrac{20,000}{7.8431391} = £2,550.$

In the United Kingdom the *effective* mortgage rate of interest is reduced by the basic rate of income tax (29% at the time of writing); thus the 12% is reduced to around *8.4%*. To obtain the *actual* mortgage payments in the UK the calculation needs to be repeated with *8.4%* instead of 12%. If you do this calculation you would obtain *£1,938*.

Exercise 8

Find the annual mortgage payments for a 20-year mortgage of £25,000 if the rate of interest is: (i) 10%; (ii) 15%.
(*Answers:* £2,936.49; £3,994.04)

DISCOUNTED CASH FLOW

A company may receive several proposals for investment projects. It is necessary for the company to have some procedure to rank investment proposals and to choose the proposal(s) which is (are) likely to prove of greatest benefit to the company. Various methods of investment appraisal are discussed in accounting and economic textbooks. In this chapter *three* methods will be considered.

(i) Payback method

This finds the number of years of net revenue required to return the initial investment.

(ii) Net present value (NPV) method

This is present value of future net revenues discounted at the appropriate rate of interest *less* the initial cost of the investment.

(iii) Internal rate of return (IRR) method

This is the rate of interest which equates the present value of future net revenues with the initial cost of the investment.

These three methods are considered in Worked Examples 9, 10 and 11.

Worked Example 9

A company has the choice of two items of capital equipment, machine A and machine B. The relevant details are shown in Table 14.1

Table 14.1

	Machine A	Machine B
Cost	£10,000	£10,000
Scrap value	£100	£100
Life	5 years	4 years
Net revenue:		
Year 1	£4,000	£5,000
2	£3,000	£4,000
3	£3,000	£3,000
4	£2,500	£2,000
5	£2,500	—

Using the *payback* method, which is the better investment?

Table 14.2

	Machine A		Machine B	
Year	Net revenue	Cumulative net revenue	Net revenue	Cumulative net revenue
1	4,000	4,000	5,000	5,000
2	3,000	7,000	4,000	9,000
3	3,000	10,000	3,000	12,000
4	2,500	12,500	2,000	14,000
5	2,500	15,000	—	—

From the calculations in Table 14.2, in the case of machine A, the cost of

£10,000 is recovered in exactly 3 years. In the case of machine B, the cost is recovered in 2.33 years. Under the *payback method* machine B would be preferred.

The payback method is easy to calculate; however, it can lead to wrong decisions:

(i)   because it ignores net revenue beyond the payback period;
(ii) because it fails to take into account the time value of money.

Exercise 9

A company is considering two alternative projects; the costs and estimated cash flows are given in Table 14.3.

Table 14.3

	Project A	Project B
Investment	£110,000	£90,000
Cash flows year 1	£60,000	−£15,000 (loss)
Cash flows year 2	£40,000	£35,000
Cash flows year 3	£30,000	£60,000
Cash flows year 4	£20,000	£80,000

Using the payback method, which project is preferred?
(*Answer:* A – pay back in less than 3 years)

Worked Example 10

Using the information in Worked Example 9, calculate the *net present value* for both machines assuming a discount rate of 10%.

The NPV method involves changing all future net revenues to present values. Suppose a project has an expected life of n years, with R_1, R_2, R_3, ..., R_n being the net revenues for years 1, 2, 3, ..., n respectively; then if i is the rate of interest and I the initial investment, the NPV is:

$$NPV = \frac{R_1}{(1+i)} + \frac{R_2}{(1+i)^2} + \frac{R_3}{(1+i)^3} + \ldots + \frac{R_n}{(1+i)^n} - I$$

As the discount rate is 10%, $(1+i) = 1.1$. The calculations are presented in Table 14.4.

Table 14.4

	Machine A		Machine B	
		− 10,000		− 10,000
Year 1	$\dfrac{4,000}{1.1}$	+ 3,636	$\dfrac{5,000}{1.1}$	+ 4,545
Year 2	$\dfrac{3,000}{1.1^2}$	+ 2,479	$\dfrac{4,000}{1.1^2}$	+ 3,306

Table 14.4 – cont

	Machine A		Machine B	
		− 10,000		− 10,000
Year 3	$\dfrac{3{,}000}{1.1^3}$	+ 2,254	$\dfrac{3{,}000}{1.1^3}$	+ 2,254
Year 4	$\dfrac{2{,}500}{1.1^4}$	+ 1,708	$\dfrac{2{,}000}{1.1^4}$	+ 1,366
Year 5	$\dfrac{2{,}500}{1.1^5}$	+ 1,552	—	—
Scrap	$\dfrac{100}{1.1^5}$	+ 62	$\dfrac{100}{1.1^4}$	+ 68
NPV		+ 1,691		+ 1,539

For *both* machines the NPV is positive, so that both machines would benefit the company. However, as machine A has the larger NPV, this machine is the better investment using the *net present value method*.

Exercise 10

Using the information in Exercise 9, calculate the NPV for both projects assuming a discount rate of: (i) 10%; (ii) 20%. (*Answers:* £13,802, − £5,216; B £25,010, − £4,892)

Worked Example 11

Using the information in Worked Example 9, calculate the internal rate of return for both machines.

Many businessmen prefer to have the return on an investment expressed as a percentage. As stated previously, the *internal rate of return* (IRR) is defined as that rate of interest which equates the present value of future net revenues with the initial cost of the investment. The equation for calculating IRR is given below:

$$= \frac{R_1}{(1+i)} + \frac{R_2}{(1+i)^2} + \frac{R_3}{(1+i)^3} + \ldots + \frac{R_n}{(1+i)^n} - I = 0$$

On comparing this equation with the formula for NPV it will be noted that the IRR is that rate of interest which makes NPV *equal to zero*.

In an examination situation candidates are usually helped to find the IRR. A typical examination question asks for NPV at (say) 10% and 20%. Suppose that at 10% the NPV is positive, and at 20% negative; then the IRR would be between 10% and 20%.

In Worked Example 10 the NPV at a 10% discount rate for machines A and B was 1,691 and 1,539 respectively. If the calculation is repeated for 20% the NPV for machines A and B is − 597 and − 307 respectively. These results are plotted in Figs. 14.1 and 14.2.

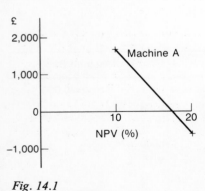

£

2,000

1,000

0

−1,000

10   20

Machine A

NPV (%)

Fig. 14.1

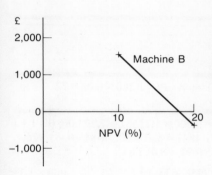

£
2,000
1,000
0
−1,000

Machine B

10 20

NPV (%)

Fig. 14.2

A line is drawn to join the two points plotted in each graph. NPV is zero where this line cuts the horizontal axis. The rate of interest at which this intersection takes place is the IRR. From the graphs it can be seen that the IRR is about 17% for machine A and 18% for machine B.

It is also possible to estimate the IRR using a *proportion* method rather than a *graphical* method. If a is NPV at X% and b is NPV at Y%, then IRR is given by the formula

$$\text{IRR} = X + \frac{a \times (Y - X)}{(a - b)}$$

In the case of machine A, $a = 1,691$, $X = 10$, $b = -597$, $Y = 20$.

Substituting, $\text{IRR} = 10 + \dfrac{1,691 \times (20 - 10)}{(1,691 - (-597))} = 10 + 7.39 = 17.4\%.$

For machine B a similar calculation gives 18.3%.

The graphical and proportion methods are approximate – but acceptable to examiners. A more accurate method using a computer gives 17.01 for machine A and 18.11 for machine B.

On the basis of the *IRR method*, machine B would be preferred.

Exercise 11

Using the information in Exercise 9 and the results of Exercise 10, estimate the IRR for each investment.
(*Answers:* 17.3; 18.4)

COMPARISON OF NPV AND IRR METHODS

The two methods effectively use the same formula. For the NPV method we calculate NPV for the rate (or rates) of interest at which the company makes its investment appraisal. In the above examples, at a rate of 10% the company would have chosen machine A; however, at 15% the NPV for machine B is the larger (547 against 441 for machine A). The NPV method is, therefore, sensitive to the *particular* rate of interest used by the company.

For Worked Examples 10 and 11 the data was carefully selected so that if an interest rate of 10% were used the NPV method selected machine A, whereas the IRR method selected machine B. In most cases the NPV and IRR rank projects in the same way.

EXAMINATION QUESTIONS

Note: In the exercises so far attempted a worked example of the same type precedes the exercise; most students can deal with compound interest problems in this situation. Students and examinees have much more difficulty with questions of the type set in the Recent Examination Questions below. Questions involving annuities and sinking funds are often done badly – candidates often find an annuity when they should be finding a sinking fund. Questions involving NPV and IRR are usually answered much more successfully.

C. SOLUTIONS TO EXERCISES

S1

(i) $i = 42/700 = 0.06$; (ii) $0.06 \times 100 = 6\%$; $Pi = 1,200 \times 0.06 = 72$

S2

(a) (i) $1 + i = 1.05$, $6,000 \times 1.05^4 = 6,000 \times 1.21551 = 7,293.06$; (ii) $1 + i = 1.1$, $6,000 \times 1.1^4 = 6,000 \times 1.4641 = 8,784.60$; (iii) $1 + i = 1.075$, $6,000 \times 1.075^4 = 6,000 \times 1.3354691 = 8,012.81$

(b) (i) $6,000 \times 1.05^{12} = 6,000 \times 1.7958563 = 10,775.14$; (ii) $6,000 \times 1.1^{12} = 6,000 \times 3.1384284 = 18,830.57$; (iii) $6,000 \times 1.075^{12} = 6,000 \times 2.3817796 = 14,290.68$

S3

(i) $APR = (1 + i)^{12} - 1 = 1.0225^{12} - 1 = 1.30605 - 1 = 0.30605 = 30.61\%$;

(ii) $APR = (1 + i)^2 - 1 = 1.14^2 - 1 = 1.2996 - 1 = 0.2996 = 29.96\%$

S4

(i) $\text{Amount} = 1,000 \times \dfrac{(1.1^6 - 1)}{0.1} = 1,000 \times \dfrac{(1.771561 - 1)}{0.1} = 1,000 \times \dfrac{0.771561}{0.1} = 7,715.61$; (ii) $\text{amount} = 1,000 \times \dfrac{(1.06^6 - 1)}{0.06} = 1,000 \times \dfrac{(1.4185191 - 1)}{0.06} = 1,000 \times \dfrac{0.4185191}{0.06} = 6,975.32$

S5

(i) $\text{Amount} = 1.1 \times 1,000 \times \dfrac{(1.1^5 - 1)}{0.1} = 1,100 \times \dfrac{(1.61051 - 1)}{0.1} = 1,100 \times \dfrac{0.61051}{0.1} = 6,715.61$; (ii) $\text{amount} = 1.07 \times 1,000 \times \dfrac{(1.07^5 - 1)}{0.07} = 1,070 \times \dfrac{(1.4025517 - 1)}{0.07} = 1,070 \times \dfrac{0.4025517}{0.07} = 6,153.29$

S6

(i) $PV = \dfrac{5,500}{1.05^5} = \dfrac{5,500}{1.27628} = 4,309$; (ii) $\dfrac{5,500}{1.07^5} = \dfrac{5,500}{1.4025517} = 3,921$

S7

(a) (i) $PV = \dfrac{2,000}{1.15} + \dfrac{2,000}{1.15^2} + \dfrac{2,000}{1.15^3} + \dfrac{2,000}{1.15^4} + \dfrac{2,000}{1.15^5} + \dfrac{2,000}{1.15^6}$

$= 1.739.13 + 1,512.29 + 1,315.03 + 1,143.51 + 994.35 + 864.66 = 7,568.97$

(ii) $PV = \dfrac{2,000}{1.07} + \dfrac{2,000}{1.07^2} + \dfrac{2,000}{1.07^3} + \dfrac{2,000}{1.07^4} + \dfrac{2,000}{1.07^5} + \dfrac{2,000}{1.07^6} = 1,869.16$

$$+ 1{,}746.88 + 1{,}632.60 + 1{,}525.79 + 1{,}425.97 + 1{,}332.68 = 9{,}533.08$$

(b)   (a) $\times (1 + i)$: (i) $7{,}568.97 \times 1.15 = 8{,}704.32$; (ii) $9{,}533.08 \times 1.07$
= 10,200.40

S8

(i) $25{,}000 = A\dfrac{(1 - (1 + i)^{-20})}{i}$, $(1 + i) = 1.1$, $1.1^{-20} = \dfrac{1}{1.1^{20}} = 0.1486436$

$25{,}000 = A\dfrac{(1 - 0.1486436)}{0.1} = A \times 8.5135637$, $A = \dfrac{25{,}000}{8.5135637}$
= 2,936.49

(ii) $(1 + i) = 1.15$, $1.15^{-20} = \dfrac{1}{1.15^{20}} = 0.0611002$

$25{,}000 = A\dfrac{(1 - 0.0611002)}{0.15} = A \times 6.2593315$, $A = \dfrac{25{,}000}{6.2593315}$
= 3,994.04

S9

| | | Project A | | Project B |
	Year	Net revenue	Cumulative net revenue	Net revenue	Cumulative net revenue
	1	60,000	60,000	−15,000	−15,000
	2	40,000	100,000	35,000	20,000
	3	30,000	130,000	60,000	80,000
	4	20,000	150,000	80,000	160,000

Project A is paid back in 2.33 years, B in 3.125 years; A is preferred.

S10

(i)		Project A		Project B
Investment		−110,000		−90,000
Year 1	$\dfrac{60{,}000}{1.1}$	54,545	$\dfrac{-15{,}000}{1.1}$	−13,636
Year 2	$\dfrac{40{,}000}{1.1^2}$	33,058	$\dfrac{35{,}000}{1.1^2}$	28,926
Year 3	$\dfrac{30{,}000}{1.1^3}$	22,539	$\dfrac{60{,}000}{1.1^3}$	45,079
Year 4	$\dfrac{20{,}000}{1.1^4}$	13,660	$\dfrac{80{,}000}{1.1^4}$	54,641
NPV		13,802		25,010

Year 1	$\dfrac{60{,}000}{1.2}$	50,000	$\dfrac{-15{,}000}{1.2}$	$-12{,}500$
Year 2	$\dfrac{40{,}000}{1.2^2}$	27,778	$\dfrac{35{,}000}{1.2^2}$	24,306
Year 3	$\dfrac{30{,}000}{1.2^3}$	17,361	$\dfrac{60{,}000}{1.2^3}$	34,722
Year 4	$\dfrac{20{,}000}{1.2^4}$	9,645	$\dfrac{80{,}000}{1.2^4}$	38,580
NPV		$-5{,}216$		$-4{,}892$

S11

$$IRR = X + \frac{a \times (Y - X)}{(a - b)}$$

For A, $a = 13{,}802$, $b = -5{,}216$, $X = 10$, $Y = 20$

$$IRR = 10 + \frac{13{,}802 \times (20 - 10)}{(13{,}802 - (-5{,}216))} = 17.3$$

For B, $a = 25{,}010$, $b = -4{,}892$, $X = 10$, $Y = 20$

$$IRR = 10 + \frac{25{,}010 \times (20 - 10)}{(25{,}010 - (-4{,}892))} = 18.4$$

D. RECENT EXAMINATION QUESTIONS

Q1

Note: Some questions included extracts of financial tables. These extracts are not included below but the solutions are based on the tables supplied. Your answers may show small arithmetic differences.

(a) Your company has decided to set up a fund for its employees with an initial payment of £2,750 which is compounded six-monthly over a four-year period at 3.5% per six months.
(i) Calculate the size of the fund to two decimal places at the end of the four years.
(ii) Calculate the effective annual interest rate, to two decimal places.
(b) The company has purchased a piece of equipment for its production department at a cost of £37,500 on 1 April 1984. It is anticipated that this piece of equipment will be replaced after five years of use on 1 April 1989. The equipment is purchased with a five-year loan, which is compounded annually at 12%.
(i) Determine the size of the equal annual payments.
(ii) Display a table which shows the amount outstanding and interest

for each year of the loan.

(c) If in (b) the £37,500 debt is compounded annually at 12% and is discharged on 1 April 1989 by using a sinking fund method, under which five equal deposits are made starting on 1 April 1984 into the fund paying 8% annually,

(i) determine the size of equal annual deposits in the sinking fund, and

(ii) display a table which demonstrates the growth of the loan and the sinking fund.

(CACA (ACCA) Dec. 1984)

(*Answers:* £3,621.22; 7.12%; £10,402.87; £10,430.64)

Q2

A manufacturing company has an ageing machine which has been requiring more maintenance recently. The directors have undertaken an investigation of a possible replacement of the machine, as they have just signed a new contract to supply parts to a large company. This contract will last for a five-year period beginning on January 1st 1985 and ending on December 31st 1989.

The alternative to replacement is the complete overhaul of the machine, but the annual maintenance costs will be higher.

The cost of a new machine is £45,000, while the cost of a complete overhaul of the old machine is £27,000; either payment being due on January 1st 1985.

The maintenance costs on either machine are paid at the end of each year of the project. The first maintenance payment for the new machine is £2,500, which is set to rise by 7.5% per annum, while that for the old machine is £4,000, which is set to rise by 10.5% per annum.

The sales of the manufactured parts from each machine are shown in Table Q2, with the inflow of funds assumed to be at the end of each year.

Table Q2

Year	1985	1986	1987	1988	1989
New machine (£)	20,000	22,000	24,000	26,000	27,000
Overhauled machine (£)	20,000	21,500	22,500	23,000	23,000

(a) Establish and tabulate the net cash flows for each plan, assuming that all other costs are the same for both.

(b) Establish the net present value of each plan assuming a 12% cost of capital.

(c) Interpret the results that you have obtained in parts (a) and (b).

(d) Without performing any further calculations, would you expect the cost of capital at which the net present value of replacement is equal to the net present value of overhauling to be above or below 12%? State your reasoning.

(CACA (ACCA) June 1985)

(*Answers:* see solutions)

Q3

Manco plc invests in a new machine at the beginning of year 1 which costs £15,000. It is hoped that the net cash flows over the next five years will correspond to those given in the table below.

Year	1	2	3	4	5
Net cash flow (£)	1,500	2,750	4,000	5,700	7,500

(a) Calculate:
  (i) The net present value assuming a 15% cost of capital.
  (ii) The net present value assuming a 10% cost of capital.
  (iii) The internal rate of return of the above project using the results of (i) and (ii).
(b) An alternative machine would cost £17,500 but would produce equal net cash flows of £5,500 over the next five years. What cost of capital would produce a break-even situation on the project?

(CACA (ACCA) Dec. 1984)

(*Answers:* −£1,998.06; £191.30; 10.4; 17.4%)

E. OUTLINE ANSWERS TO EXAM QUESTIONS

A1

(a) (i) $P_n = P_0 \times (1+i)^8$, $P_n = 2,750 \times 1.035^8 = 2,750 \times 1.316809 = 3,621.22$; (ii) $((1+i)^2 - 1) = 1.035^2 - 1 = 1.071225 - 1 = 0.0712 = 7.12\%$

(b) The question does not state when payments are made; the solution assumes that payments are at the end of the year.

(i) The present value of the annual payments must equal £37,500.

$$37,500 = \frac{A}{1.12} + \frac{A}{1.12^2} + \frac{A}{1.12^3} + \frac{A}{1.12^4} + \frac{A}{1.12^5}$$

$$37,500 = A(0.892857 + 0.797194 + 0.711780 + 0.635518 + 0.567427)$$

$$37,500 = A \times 3.604776, \quad A = \frac{37,500}{3.604776} = 10,402.87$$

(ii)	Amount on 1 April	Interest	Payment
1984	37,500.00	4,500.00	10,402.87
1985	31,597.13	3,791.66	10,402.87
1986	24,985.92	2,998.31	10,402.87
1987	17,581.36	2,109.76	10,402.87
1988	9,288.25	1,115.59	10,402.87
1989	−0.03*		

*Result not zero due to rounding.

(c) Debt 1 April 1989, $37,500 \times 1.12^5 = 37,500 \times 1.7623417$
$= 66,087.81$. If A is amount, $A \times (1.08 + 1.08^2 + 1.08^3 + 1.08^4 + 1.08^5) = 66,087.81$. $A(1.08 + 1.1664 + 1.259712 + 1.360489 + 1.469328) = A \times 6.335929$.

$$A = \frac{66,087.81}{6.335929} = 10,430.64$$

1 April	Loan	Payment	Sinking Fund	Interest
1984	37,500	10,430.64	11,265.09	834.45
1985	42,000	10,430.64	21,695.73	1,735.66
1986	47,040	10,430.64	33,862.03	2,708.96
1987	52,684.80	10,430.64	47.001.63	3,760.13
1988	59,006.98	10,430.64	61,192.40	4,895.39
1989	66,087.81		66,087.79	

A2

(a)

End of	Machine replacement Cash outflow	Cash inflow	Net cash inflow	Machine overhaul Cash outflow	Cash inflow	Net cash inflow
1984	45,000.00	—	−45,000.00	27,000.00	—	−27,000.00
1985	2,500.00	20,000	17,500.00	4,000.00	20,000	16,000.00
1986	2,687.50	22,000	19,312.50	4,420.00	21,500	17,080.00
1987	2,889.06	24,000	21,110.94	4,884.10	22,500	17,615.90
1988	3,105.74	26,000	22,894.26	5,396.93	23,000	17,603.07
1989	3,338.67	27,000	23,661.33	5,963.61	23,000	17,036.39

(b)

End of year	Discount factor	Machine replacement Net cash flow	Present value	Machine overhaul Net cash flow	Present value
1984	1.0000	−45,000.00	−45,000.00	−27,000.00	−27,000.00
1985	0.8929	17,500.00	15,625.75	16,000.00	14,286.40
1986	0.7972	19,312.50	15,395.93	17,080.00	13,616.18
1987	0.7118	21,110.94	15,026.77	17,615.90	12,539.00
1988	0.6355	22,894.26	14,549.30	17,603.07	11,186.75
1989	0.5674	23,661.33	13,425.44	17,036.39	9,666.45
NPV			29,023.19		33,294.78

(c) NPV of overhaul greater than NPV of new, thus overhaul preferred. A new machine, however, has a higher cash inflow.

(d) As the cash inflows for new exceed the cash inflows for overhaul at zero discounting, but is reversed at 12%; equality occurs at less than 12%.

(a) (i), (ii)

Year	Cash flow	Discount factor (15%)	Present value	Discount factor (10%)	Present value
0	− 15,000	1.0000	− 15,000.00	1.0000	− 15,000.00
1	1,500	0.8696	1,304.40	0.9091	1,363.65
2	2,750	0.7561	2,079.28	0.8264	2,272.60
3	4,000	0.6575	2,630.00	0.7513	3,005.20
4	5,700	0.5718	3,259.26	0.6830	3,893.10
5	7,500	0.4972	3,729.00	0.6209	4,656.75
NPV			− 1,998.06		191.30

(iii) $\text{IRR} = X + \dfrac{a \times (Y - X)}{(a - b)}$, $a = 191.30$, $b = -1,998.06$, $Y = 15$, $X = 10$

$$\text{IRR} = 10 + \frac{191.30 \times (15 - 10)}{(191.30 - (-1,998.06))} = 10.4$$

(b) This problem requires a trial-and-error solution to find IRR. If you try 15% discount rate the NPV is found to be £937.10, at zero rate of discount the NPV is £6,450; the IRR is, therefore, more than 15%. At 20% NPV is found to be − £1,052.25. Using the formula IRR is found to be 17.4%.

A STEP FURTHER

Tennant-Smith, *Mathematics for the Manager*, Ch. 7.

Chapter 15   Linear programming

A. GETTING STARTED

In business and industry we frequently meet the problem of how to use limited resources to best advantage. For example, in a factory we might wish to maximize profit, yet we will inevitably face a number of constraints – the amount of floor space available; the number of machines; the amount of capital available to purchase more and perhaps better machines; the number of skilled mechanics that can be recruited, etc.

In Chapter 12, in Exam Question 5, we found the number of units to make of each product so that all the resources available were fully utilized. However, the solution found to that problem is not necessarily the utilization of resources which *maximizes profits*. If a firm makes several products and the profit per item of one product is very much greater than the profit per item of the other products, it may be necessary to *under-utilize* some resources in order to make more of the highly profitable product, and in this way to maximize profits. In such situations the technique of *linear programming* enables us to find the optimum use of the resources available.

B. ESSENTIAL PRINCIPLES

The technique is called linear programming because it assumes linearity, i.e. if 5 men produce 100 items per hour, 20 men would produce 400 items per hour. No provision is made for economies of scale – in more advanced texts, the technique of non-linear programming is explained.

The methods of linear programming are best explained by an example.

Worked Example 1

A manufacturer makes two products X and Y. Each product has to be processed in three departments: mechanical, electrical and assembly.

Each unit of X requires 12 minutes in the mechanical department, 6 minutes in the electrical department and 3 minutes in the assembly department. The corresponding times for each unit of Y are 6, 9 and 3 minutes respectively. In a given period of time there are 120 hours available in the mechanical department, 90 hours available in the electrical department and 35 hours available in the assembly department. For the product Y the selling price is £8 and the cost of materials and labour is £5. The corresponding amounts for product X are £12 and £7 respectively.

Find how many items of X and Y should be produced to maximize the contribution to profits and fixed overheads.

A useful procedure is to set out the information as in Table 15.1.

Table 15.1

	Product		Constraint in minutes
	X	Y	
Time to produce one unit:			
Mechanical	12 min	6 min	$120 \times 60 = 7{,}200$
Electrical	6 min	9 min	$90 \times 60 = 5{,}400$
Assembly	3 min	3 min	$35 \times 60 = 2{,}100$
Price	£12	£8	
Cost of materials and labour	£7	£5	
	—	—	
Contribution to profit and fixed overheads	£5	£3	

Note: It is essential to work in the same units. The constraints given in hours in the question have been changed to minutes.

Suppose we make X units of product X, and Y units of product Y. As each unit of X requires 12 minutes in the *mechanical department*, then X units of product X will require $12X$ minutes. Similarly as each unit of Y requires 6 minutes in mechanical department, then Y units of product Y will require $6Y$ minutes. Thus the total time in the mechanical department to make X units of product X, and Y units of product Y, is $12X + 6Y$. The total time available is 7,200 minutes; hence $12X + 6Y$ *cannot exceed* 7,200 minutes. Thus we have the inequality:

$$12X + 6Y \leqslant 7{,}200$$

Likewise the inequalities for the *electrical* and *assembly* departments are:

$$6X + 9Y \leqslant 5{,}400$$
$$3X + 3Y \leqslant 2{,}100$$

Now we cannot produce negative quantities of products X and Y; thus

$$X \geqslant 0 \qquad Y \geqslant 0$$

These are usually called the *non-negativity* constraints.

We are asked to maximize the contribution to profit and fixed overheads. The contribution of each unit of X is £5, thus the contribution of X units of product X is £5X. Likewise the contribution of Y units of product Y is £3Y. The *total contribution C* is equal to $5X + 3Y$. We wish to maximize:

$$C = 5X + 3Y$$

$C = 5X + 3Y$ is called the *objective function*.

If you are asked to 'formulate as a linear programming problem' the information should be set out as follows:

Maximize	$C = 5X + 3Y$	
Subject to	$12X + 6Y \leqslant 7,200$	(1)
	$6X + 9Y \leqslant 5,400$	(2)
	$3X + 3Y \leqslant 2,100$	(3)
	$X \geqslant 0$	(4)
	$Y \geqslant 0$	(5)

Note: Examination candidates often (a) omit the non-negativity constraints; (b) have an equals sign ($=$) instead of an inequality sign ($\leqslant$). These errors lose marks.

Values of X and Y which satisfy all the constraints (1) to (5) are called a 'feasible solution'. We need to find the feasible solution which maximizes $C = 5X + 3Y$.

GRAPHICAL SOLUTION

Let us take the first constraint $12X + 6Y \leqslant 7,200$. We plot the line $12X + 6Y = 7,200$. To plot this line the easiest procedure is to note that when $X = 0$, $Y = 1,200$, and when $Y = 0$, $X = 600$. Plot these two points and use a ruler and a sharp pointed pencil to join the two points. This is shown in Fig. 15.1.

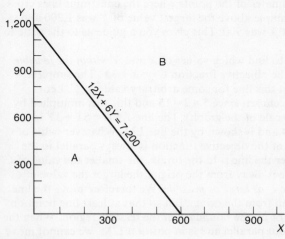

Fig. 15.1   *Fig. 15.2*

Any value of X and Y in area A or on the line $12X + 6Y = 7,200$ satisfies the inequality $12X + 6Y \leqslant 7,200$. Values of X and Y in area B *do not* satisfy the inequality. Let us plot on the same graph the line $6X + 9Y = 5,400$. We note that when $X = 0$, $Y = 600$, and when $Y = 0$, $X = 900$. This line is plotted in Fig. 15.2. Any value of X and Y in area C satisfies constraints (1), (2), (4) and (5). Let us now plot the line $3X + 3Y = 2,100$. We note that when $X = 0$, $Y = 700$, and when $Y = 0$, $X = 700$. This line is plotted in Fig. 15.3. Any value of X and Y in the shaded area satisfies constraints (1) to (5). This area is called the *feasible region*.

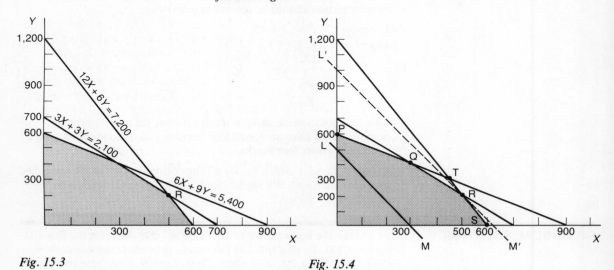

Fig. 15.3   *Fig. 15.4*

In order to obtain the correct scales for your graph, the first step is to find the coordinates of the points where the constraint lines cross the axes. In the example above the largest value of Y was 1,200 and the largest value of X was 900. This gives you a guide as to the scale to use.

We now wish to find which values of X and Y *within the feasible region* maximizes the objective function $C = 5X + 3Y$. The simplest procedure is to plot this line for some arbitrary value of C. Let $C = 1,500$ (1,500 was chosen since $5 \times 3 = 15$ and this was multiplied by 100 in view of the scale of the graph). The line $1,500 = 5X + 3Y$ is plotted in Fig. 15.4 and is shown by the line LM. Whatever value of C we choose the line of the objective function is always parallel to the line LM. The nearer the line is to the origin, the smaller the value of C. The further the line is away from the origin, the larger the value of C. Our aim is to make *C as large as possible*. We therefore move the line LM as far as we can from the origin, but so that at least one point on the line lies within or on the boundary of the feasible region. When the line LM is moved out parallel and is in position L'M' we cannot move the line any further because we would be entirely outside the feasible region. The coordinates of the point R give the values of X and Y which maximize C. We note that the coordinates of R are $X = 500$, $Y = 200$. Thus if we make 500 of product X and 200 of product Y we maximize the contribution C, and the maximum contribution C is $5 \times 500 + 3 \times 200 = £3,100$.

Notes on solution:

(i) The solution to a linear programming problem is at one of the *vertices* of the feasible region, i.e. at either P, Q, R or S – these are marked in Fig. 15.4. The **gradient** of the objective function enables us to decide which vertex gives the maximum.

(ii) It is possible to find the coordinates of P, Q, R, and S by solving simultaneous equations. The value of C can then be found by substituting the coordinates for each point into the objective function $C = 5X + 3Y$. We can thus determine which vertex gives maximum contribution. This method is **not** recommended since it is so easy to make a mistake. Some candidates are likely to find the coordinates of point T (see Fig. 15.4) which gives the largest contribution but is outside the feasible region.

(iii) If R is located by the graphical method, it is not always easy to read off the coordinates accurately; by noting that R is at the intersection of the lines $3X + 3Y = 2,100$ and $12X + 6Y = 7,200$ (see Fig. 15.3) we can also solve the simultaneous equations for X and Y. This is a useful check on the values of X and Y given by the graphical method.

(iv) It will be noted that the point R is at the intersection of the constraint lines for *mechanical* and *assembly* departments. This means that these departments are operating at full capacity. The electrical department, however, has spare capacity. At a production level of 200 of Y and 500 of X this takes $6 \times 500 + 9 \times 200 = 4,800$ minutes in the electrical department; thus $5,400 - 4,800 = 600$ minutes $= 10$ hours are still available in the *electrical* department.

Exercise 1

A company manufactures two products X and Y. Each product has to be processed in three departments: welding, assembly and painting. Each unit of X spends 2 hours in the welding department, 3 hours in assembly and 1 hour in painting. The corresponding times for a unit of Y are 3, 2 and 1 respectively. The man-hours available in a month are 1,500 for the welding department, 1,500 in assembly and 550 in painting. The contribution to profits and fixed overheads are £100 for product X and £120 for product Y.

(i) Formulate as a linear programming problem.

(ii) Represent graphically and shade feasible region.

(iii) Obtain the optimal solution and find the maximum contribution.

(iv) Which department has spare capacity and how much?

(*Answers:* Y = 400, X = 150; 63,000; Assembly, 4 hours 10 minutes)

OTHER CONSTRAINTS

Worked Example 2

In Worked Example 1 there were three constraints (ignoring non-negativity). It would be possible to include additional constraints. Suppose that the manufacturer had a contract to supply a customer with 300 items of product Y. This imposes the constraint $Y \geqslant 300$. The line $Y = 300$ is *horizontal* and has been drawn in Fig. 15.5. The new feasible region is the shaded area. The optimal production level is now the point T, when the objective function is in position L'M'. The coordinates of T are $X = 400$, $Y = 300$.

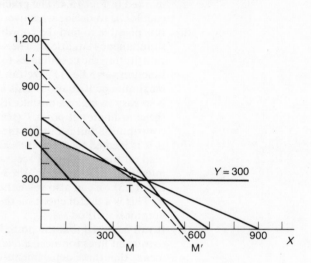

Fig. 15.5

There may be other constraints. If there was a *limit* on the number of X that could be sold, say not more than 600, this would mean a constraint $X \leqslant 600$ and would be represented by a *vertical* line through $X = 600$. Alternatively, suppose that the company wished to sell *at least* 600 of the products, this would mean that $X + Y \geqslant 600$.

Exercise 2

A company manufactures two products X and Y. The production facilities restrict production to a total of 50 units per day. Each day, 20 man-hours are available in the assembly shop, and 32 man-hours in the paint shop. Each unit of X requires 30 minutes in the assembly shop and 24 minutes in the paint shop. The corresponding times for product Y are 10 minutes and 48 minutes respectively. The contribution to fixed overheads and profits for each unit of product X is £9, and for a unit of product Y is £12.

Advise the company on the optimal product mix. Find the maximum contribution.

(*Answers:* $Y = 30$, $X = 20$; 540)

MINIMIZATION

Linear programming may be used to solve minimization problems, for example, how to produce the required output at minimum cost. The method may be illustrated by a worked example.

Worked Example 3

A farmer uses two feedstuffs Y and X to feed his animals. To obtain meat of good quality he has to ensure that he obtains each day 240 kilos of fat, 150 kilos of protein and 30 kilos of vitamins. Feed Y contains by weight 30% fat, 10% protein and 6% vitamins. Feed X contains by weight, 20% fat, 25% protein and 2% vitamins. The cost per kilo of feed Y is 30 pence and feed X is 15 pence. Find the minimum cost quantities of feed Y and X to purchase each day to meet the needs of the animals.

Table 15.2

	Feed		Daily requirements (Kilos)
	Y	X	
Fat	30%	20%	240
Protein	10%	25%	150
Vitamins	6%	2%	30
Cost per kilo (pence)	30	15	

The information is set out in Table 15.2. Suppose Y kilos is the amount of feed Y to purchase, and X kilos the amount of feed X to purchase, then $0.30Y + 0.20X \geqslant 240$ is the *fat constraint* since at least 240 kilos of fat are required. The other constraints are:

$0.10Y + 0.25X \geqslant 150$ protein constraint
$0.06Y + 0.02X \geqslant 30$ vitamin constraint

In addition there are the non-negativity constraints. We need to minimize the costs, i.e. minimize $C = 30Y + 15X$.

The formulation should be set out as follows:

Minimize $C = 30Y + 15X$

Subject to $30Y + 20X \geqslant 24{,}000$ (1)

                 $10Y + 25X \geqslant 15{,}000$ (2)

                 $6Y + 2X \geqslant 3{,}000$ (3)

                 $X \geqslant 0$ (4)

                 $Y \geqslant 0$ (5)

Constraints (1) to (3) were simplified by multiplying by 100 to clear decimals. There is no need to do this; Fig. 15.6 can be obtained from either form of the equation.

The feasible region is shaded. The objective function $C = 30Y + 15X$ is plotted at LM ($C = 24{,}000$ was used). The object is to *minimize C*. We

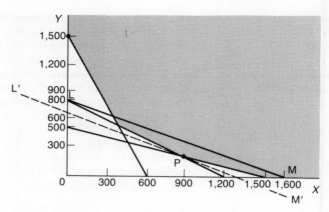

Fig. 15.6

thus need to have the line LM as *near as possible to the origin* but within the feasible region. The position of this line is L'M' and the optimal solution is given by point P, with $X = 900$, $Y = 200$.

Thus the optimal solution is to buy 900 kilos of feed X and 200 kilos of feed Y at a cost of $30 \times 200 + 15 \times 900 = 19,500 = £195$.

| Exercise 3 | To obtain maximum yield from a field, a farmer requires at least 300 units of nitrate, 240 units of potash and 90 units of phosphorus. He can buy two compound fertilizers A and B. Each kilo of fertilizer A contains 10 units of nitrates, 5 units of potash, and 6 units of phosphorus; the corresponding number of units for fertilizer B are 5, 10 and 1 respectively. If the cost of fertilizer A is £1.50 per kilo and £2.00 per kilo for fertilizer B, find the minimum cost quantities to purchase. What is the cost and which chemical is obtained at more than the minimum quantities and by what amount? (*Answers: A = 24, B = 12; 60; phosphorus; 66 units*) |

EXAMINATION QUESTIONS — **Note:** Most examination questions set at the introductory level involve maximization; questions are rarely set on minimization. The Exam Questions include one or two variations on the standard questions.

C. SOLUTIONS TO EXERCISES

S1

Maximize $C = 100X + 120Y$
Subject to $2X + 3Y \leqslant 1,500$
             $3X + 2Y \leqslant 1,500$
             $X + Y \leqslant 550$
             $X \geqslant 0 \qquad Y \geqslant 0$

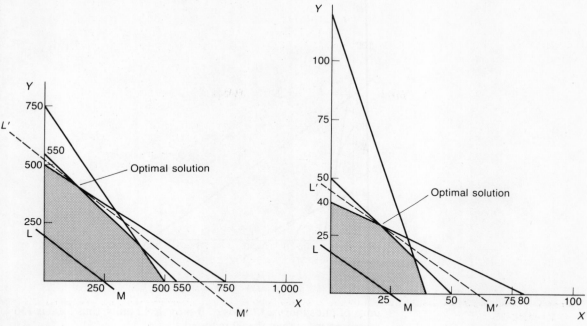

Fig. 15S.1

Fig. 15S.2

From Fig. 15S.1, the optimal solution is when $Y = 400$, $X = 150$.
$C = 100 \times 150 + 120 \times 400 = 63,000$.

The constraints for painting and welding give optimal solution; thus assembly has spare capacity $= 1,500 - (3 \times 150 + 2 \times 400) = 250$ minutes $= 4$ hours 10 minutes.

S2

Maximize	$C = 9X + 12Y$
Subject to	$30X + 10Y \leqslant 1,200$
	$24X + 48Y \leqslant 1,920$
	$X + Y \leqslant 50$
	$X \geqslant 0 \qquad Y \geqslant 0$

From Fig. 15S.2, the optimal solution is when $Y = 30$, $X = 20$.
$C = 9 \times 20 + 12 \times 30 = 540$.

S3

Minimize	$C = 1.5A + 2.0B$
Subject to	$10A + 5B \geqslant 300$
	$5A + 10B \geqslant 240$
	$6A + B \geqslant 90$
	$A \geqslant 0 \qquad B \geqslant 0$

From Fig. 15S.3, optimal solution is $A = 24$, $B = 12$. $C = 1.5 \times 24 + 2.0 \times 12 = 60$.

Constraints are for nitrates and potash; 24 kilos of A provide 144

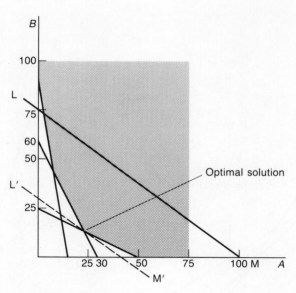

Fig. 15S.3

units of phosphorus, 12 kilos of B provide 12 units, thus total is 156 units – 66 more than the 90 required.

D. RECENT EXAMINATION QUESTIONS

Q1

A firm makes two products, X and Y. Each product has to be processed in three departments, mechanical, electrical and assembly. Each unit of X requires 8 hours in the mechanical department, 4 hours in the electrical department and 7 hours in the assembly department. The corresponding number of hours for each unit of Y are 8, 2, and 2 respectively. The firm works an 8-hour day. 7 men are employed in the mechanical department, 2 in the electrical and 3 in the assembly. The selling price is £900 for X and £600 for Y. The variable costs for each unit of X and Y are £400 and £300 respectively.

(i) Formulate as a linear programming problem.

(ii) Find the level of output of X and Y which will maximize contribution to profits and fixed overheads.

(iii) Which department is under-utilized and how many man-hours are available each day?

(*Answers: Y = 6, X = 1; assembly, 5*)

Q2

A company needs to purchase a number of small printing presses, of which there are two types, X and Y.

Type X costs £4,000, requires two operators and occupies 20 square metres of floor space. Type Y costs £12,000, also requires two operators but occupies 30 square metres.

The company has budgeted for a maximum expenditure on these presses of £120,000. The print shop has 480 square metres of available floor space, and work must be provided for at least 24 operators.

It is proposed to buy a combination of presses X and Y that will maximize production, given that type X can print 150 sheets per minute and type Y, 300 per minute.

(a) Write down all the equations/inequalities which represent the cost and space conditions. The labour conditions are given by $2X + 2Y \geqslant 24$;

(b) draw a graph to represent this problem, shading any unwanted regions;

(c) use the graph to find the numbers of presses X and Y the company should buy to achieve its objective of maximum production;

(d) state the figure of maximum production and the total cost of presses in this case.

(ICMA May 1984)

(*Answers:* $Y = 4$, $X = 18$; 3,900, £120,000)

Q3

A company manufactures two products A and B. The company has a limited budget of £8,000 to meet production costs. The production cost for each unit of product A is £16, the corresponding cost for product B is £8. For product A the time of assembly is 4 minutes, for product B the corresponding time is 12 minutes. Each week 80 man-hours are available for assembly. The packaging time for either product is 6 minutes. Each week 60 man-hours are available for packaging. The contribution to fixed overheads and profits for each product A sold is £5 and for each product B is £8.

(a) Formulate as a linear programming problem.

(b) Represent graphically and shade feasible region.

(c) Find the number of products A and B need to sell to maximize contribution. Find also the maximum contribution.

(d) If the company were forced to reduce the price of the product B so that the contribution was reduced to £4, what effect would this have on the total contribution?

(ICSA (Pilot paper))

(*Answers:* $A = 300$, $B = 300$; 3,900; $A = 400$, $B = 200$; reduction, 1,100)

Q4

A chocolate manufacturer produces two kinds of chocolate bar, X and Y, which are made in three stages: blending, baking and packaging.

The time, in minutes, required for each box of chocolate bars is as in Table Q4.

Table Q4

	Blending	Baking	Packaging
X	3	5	1
Y	1	4	3

The blending and packaging equipment is available for 15 machine-hours and the baking equipment is available for 30 machine-hours.

The contribution on each box of X is £1 and on each box of Y, £2. The machine time may be used for either X or Y at all times it is available. All production may be sold.

(a) State the equations/inequalities which describe the production conditions;

(b) draw a graph of these equations/inequalities and hence find how many boxes of each chocolate bar the manufacturer should produce to maximize contribution;

(c) state this maximum contribution and comment on your answer.

(ICMA Nov. 1984)

(*Answer:* $Y = 245$, $X = 164$; £654)

OUTLINE ANSWERS TO EXAM QUESTIONS

A1

Maximize $\quad C = 500X + 300Y$

Subject to $\quad 8X + 8Y \leqslant 56$

$\qquad\qquad\quad 4X + 2Y \leqslant 16$

$\qquad\qquad\quad 7X + 2Y \leqslant 24$

$\qquad\qquad\quad X \geqslant 0 \qquad Y \geqslant 0$

From Fig. 15S.4, optimal solution is $Y = 6$, $X = 1$; $C = 2{,}300$.

Constraints are for mechanical and electrical departments; assembly is under-utilized, $7 \times 1 + 2 \times 6 = 19$ used, thus 5 are available.

A2

Maximize $\quad P = 150X + 300Y$

Subject to $\quad 2X + 2Y \geqslant 24$

$\qquad\qquad\quad 4{,}000X + 12{,}000Y \leqslant 120{,}000$

$\qquad\qquad\quad 20X + 30Y \leqslant 480$

$\qquad\qquad\quad X \geqslant 0 \qquad Y \geqslant 0$

From Fig. 15S.5, optimal solution is $Y = 4$, $X = 18$, $P = 3{,}900$. Total cost = £120,000.

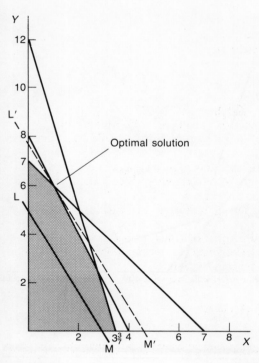

Fig. 15S.4

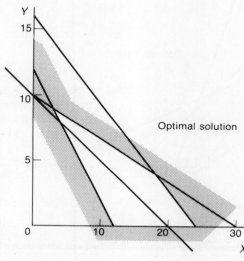

Fig. 15S.5

A3

Maximize	$C = 5A + 8B$
Subject to	$16A + 8B \leqslant 8,000$
	$4A + 12B \leqslant 4,800$
	$6A + 6B \leqslant 3,600$
	$A \geqslant 0 \qquad B \geqslant 0$

From Fig. 15S.6, optimal solution is $A = 300$, $B = 300$, $C = 3,900$.

If the price of product B is reduced to £4, the objective function is now $C = 5A + 4B$; optimal solution is now $A = 400$, $B = 200$ and $C = 2,800$. Thus contribution is reduced from 3,900 to 2,800 = 1,100.

A4

Maximize	$C = X + 2Y$
Subject to	$3X + Y \leqslant 900$
	$5X + 4Y \leqslant 1,800$
	$X + 3Y \leqslant 900$
	$X \geqslant 0 \qquad Y \geqslant 0$

From Fig. 15S.7, optimal solution is $X = 164$, $Y = 245$; $C = 654$.

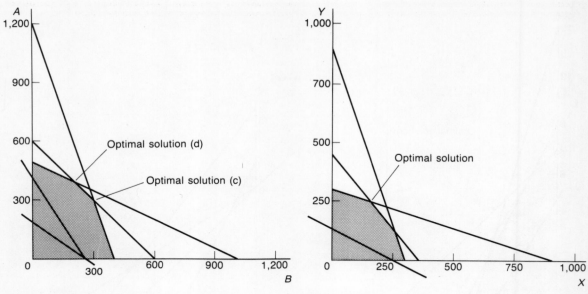

Fig. 15S.6

Fig. 15S.7

A STEP FURTHER   Tennant-Smith, *Mathematics for the Manager*, Ch. 3.

Financial tables

1 Compound interest

Year	5%	10%	15%	20%
0	1.00000	1.00000	1.00000	1.00000
1	1.05000	1.10000	1.15000	1.20000
2	1.10250	1.21000	1.32250	1.44000
3	1.15763	1.33100	1.52088	1.72800
4	1.21551	1.46410	1.74901	2.07360
5	1.27628	1.61051	2.01136	2.48832
6	1.34010	1.77156	2.31306	2.98598

2 Discount factors

Year	5%	10%	15%	20%
0	1.00000	1.00000	1.00000	1.00000
1	0.95238	0.90909	0.86957	0.83333
2	0.90703	0.82645	0.75614	0.69444
3	0.86384	0.75131	0.65752	0.57870
4	0.82270	0.68301	0.57175	0.48225
5	0.78353	0.62092	0.49718	0.40188
6	0.74622	0.56447	0.43233	0.33490

3 $S_{\overline{n}|}$ = Future value of an annuity

Year	5%	10%	15%	20%
1	1.00000	1.00000	1.00000	1.00000
2	2.05000	2.10000	2.15000	2.20000
3	3.15250	3.31000	3.47250	3.64000
4	4.31013	4.64100	4.99338	5.36800
5	5.52563	6.10510	6.74238	7.44160
6	6.80191	7.71561	8.75374	9.92992

4 $a_{\overline{n}|}$ = Present value of an annuity

Year	5%	10%	15%	20%
1	0.95238	0.90909	0.86957	0.83333
2	1.85941	1.73554	1.62571	1.52778
3	2.72325	2.48685	2.28323	2.10648
4	3.54595	3.16987	2.85498	2.58873
5	4.32948	3.79079	3.35216	2.99061
6	5.07569	4.35526	3.78448	3.32551

Normal distribution table

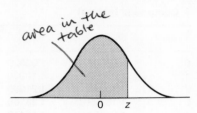

Fig. A.1

z	.00	.01	.02	.03	.04	.05	.06	.07	.08	.09
0.0	.5000	.5040	.5080	.5120	.5160	.5199	.5239	.5279	.5319	.5359
0.1	.5398	.5438	.5478	.5517	.5557	.5596	.5636	.5675	.5714	.5753
0.2	.5793	.5832	.5871	.5910	.5948	.5987	.6026	.6064	.6103	.6141
0.3	.6179	.6217	.6255	.6293	.6331	.6368	.6406	.6443	.6480	.6517
0.4	.6554	.6591	.6628	.6664	.6700	.6736	.6772	.6808	.6844	.6879
0.5	.6915	.6950	.6985	.7019	.7054	.7088	.7123	.7157	.7190	.7224
0.6	.7275	.7291	.7324	.7357	.7389	.7422	.7454	.7486	.7517	.7549
0.7	.7580	.7611	.7642	.7673	.7704	.7734	.7764	.7794	.7823	.7852
0.8	.7881	.7901	.7939	.7967	.7995	.8023	.8051	.8078	.8106	.8133
0.9	.8159	.8186	.8212	.8238	.8264	.8289	.8315	.8340	.8365	.8389
1.0	.8413	.8438	.8461	.8485	.8508	.8531	.8554	.8577	.8599	.8621
1.1	.8643	.8665	.8686	.8708	.8729	.8749	.8770	.8790	.8810	.8830
1.2	.8849	.8869	.8888	.8907	.8925	.8944	.8962	.8980	.8997	.9015
1.3	.9032	.9049	.9066	.9082	.9099	.9115	.9131	.9147	.9162	.9177
1.4	.9192	.9207	.9222	.9236	.9251	.9265	.9279	.9292	.9306	.9319
1.5	.9332	.9345	.9357	.9370	.9382	.9394	.9406	.9418	.9429	.9441
1.6	.9452	.9463	.9474	.9484	.9495	.9505	.9515	.9525	.9535	.9545
1.7	.9554	.9564	.9573	.9582	.9591	.9599	.9608	.9616	.9625	.9633
1.8	.9641	.9649	.9656	.9664	.9671	.9678	.9686	.9693	.9699	.9706
1.9	.9713	.9719	.9726	.9732	.9738	.9744	.9750	.9756	.9761	.9767
2.0	.9772	.9778	.9783	.9788	.9793	.9798	.9803	.9808	.9812	.9817
2.1	.9821	.9826	.9830	.9834	.9838	.9842	.9846	.9850	.9854	.9857
2.2	.9861	.9864	.9868	.9871	.9875	.9878	.9881	.9884	.9887	.9890
2.3	.9893	.9896	.9898	.9901	.9904	.9906	.9909	.9911	.9913	.9916
2.4	.9918	.9920	.9922	.9925	.9927	.9929	.9931	.9932	.9934	.9936
2.5	.9938	.9940	.9941	.9943	.9945	.9946	.9948	.9949	.9951	.9952
2.6	.9953	.9955	.9956	.9957	.9959	.9960	.9961	.9962	.9963	.9964
2.7	.9965	.9966	.9967	.9968	.9969	.9970	.9971	.9972	.9973	.9974
2.8	.9974	.9975	.9976	.9977	.9977	.9978	.9979	.9979	.9980	.9981
2.9	.9981	.9982	.9983	.9983	.9984	.9984	.9985	.9985	.9986	.9986
3.0	.9987	.9987	.9987	.9988	.9988	.9989	.9989	.9989	.9990	.9990

Index